Color Tab Index to Bat Genera

Large bat with folds of skin across face. Seen in deserts or near rivers. **Uncommon** ...ps

Large-eyed, leaf-nosed bats. Some found at flowers. **Uncommon; limited to Southwest or Florida Keys.** ...ris, *Leptonycteris, Artibeus*

Bat with pale yellow fur; large ears. Often feeds on ground. **Common; only in West.** *Antrozous*

Often in buildings or bat houses; very common bat in urban areas. **Common.** *Eptesicus*

Bats with very large ears. May make peeps or clicks in flight. **Uncommon; only in West.** *Euderma, Idionycteris*

Heavily furred bats; many brightly colored. Often alone in trees. **Common (some species).** *Lasionycteris, Lasiurus*

Small brownish bats. Zigzagging flight. Usually in colonies. **Common (some species).** *Myotis*

Small to smallest North American bats. Found roosting in colonies. **Common.** *Nycticeius, Pipistrellus*

Medium-sized bats; large ears. Agile fliers; can almost hover. **Uncommon (some species).** *Corynorhinus*

Large round ears bend forward; long thin tail; long pointed wings. Some give piercing call in flight. **Uncommon (some species); limited to Southwest or Florida.** *Eumops, Molossus, Nyctinomops*

Lives in urban areas, bat houses. Long pointed wings. Large colonial roosts have musty odor. **Common across southern United States.** *Tadarida*

STOKES FIELD GUIDES

Stokes Field Guide to Birds: Eastern Region

Stokes Field Guide to Birds: Western Region

Stokes Field Guide to Bird Songs: Eastern Region (CD/cassette)

Stokes Field Guide to Bird Songs: Western Region (CD/cassette)

STOKES BEGINNER'S GUIDES

Stokes Beginner's Guide to Bats

Stokes Beginner's Guide to Birds: Eastern Region

Stokes Beginner's Guide to Birds: Western Region

Stokes Beginner's Guide to Butterflies

Stokes Beginner's Guide to Dragonflies

Stokes Beginner's Guide to Shorebirds

STOKES BACKYARD NATURE BOOKS

Stokes Bird Feeder Book

Stokes Bird Gardening Book

Stokes Birdhouse Book

Stokes Bluebird Book

Stokes Butterfly Book

Stokes Hummingbird Book

Stokes Oriole Book

Stokes Purple Martin Book

Stokes Wildflower Book: East of the Rockies

Stokes Wildflower Book: From the Rockies West

STOKES NATURE GUIDES

Stokes Guide to Amphibians and Reptiles

Stokes Guide to Animal Tracking and Behavior

Stokes Guide to Bird Behavior, Volume 1

Stokes Guide to Bird Behavior, Volume 2

Stokes Guide to Bird Behavior, Volume 3

Stokes Guide to Enjoying Wildflowers

Stokes Guide to Nature in Winter

Stokes Guide to Observing Insect Lives

OTHER STOKES BOOKS

The Natural History of Wild Shrubs and Vines

Stokes
Beginner's Guide
to Bats

Kim Williams, Rob Mies *with Donald and Lillian Stokes*

Little, Brown and Company
Boston New York London

599.4
WIL

First Edition

10 9 8 7 6 5 4 3 2 1

Printed in Singapore

Library of Congress Cataloging-in-Publication Data

Stokes beginner's guide to bats / Kim Williams . . .
[et al.]. — 1st ed.
     p.   cm.
   ISBN 0-316-81658-2
   1. Bats — Identification. 2. Bats — North
America — Identification. I. Williams, Kim,
1966– II. Title.

   QL737.C5 S744 2002
   599.4 — dc21   2001038112

Photo Credits

The letter following each page number refers to
the position of the photograph on the page
(L=left; R=right).

Scott Altenbach, University of New Mexico: 3,
     43R, 47R, 49R, 51, 53, 61, 65, 71, 77, 85, 87,
     107L, 113L, 121, 123, 125, 131L
Roger W. Barbour: 45, 55R, 63, 79, 81, 93, 101,
     103, 105, 115R, 119, 127, 129
Tim Carter, Tim Carter Photography: 67R, 97,
     99L, 109, 113R
R. K. LaVal: 43L, 89

Joe McDonald, McDonald Wildlife Photography,
     Inc.: 17, 22, 95R, 111, 160
Mary Ann McDonald, McDonald Wildlife
     Photography, Inc.: 95L
Dr. Les Meade, Morehead State University: 91,
     99R, 115L, 117L, 117R
Rob Mies, Organization for Bat Conservation: 35
Bruce Montagne, Bruce Montagne
     Photography: 18, 57, 69R, 73
Charles Rau, CSR Nature Photography: 47L, 49L,
     69L, 83, 107R
Carl R. Sams II, Carl R. Sams II Photography:
     28, 75
Bruce Thomson, Queensland Parks and Wildlife
     Service: 55L, 59, 67L, 131R
Kim Williams, Organization for Bat
     Conservation: 16, 20, 25, 26, 31

Contents

Foreword

Bats capture our imagination for many reasons. First, they are active mostly at night, so we literally do not know much about them. Second, they are mammals, like us, but they can fly. Third, they often roost during the day in dark caves or crevices where we are reluctant to go. And finally, any animal that is more comfortable in the night than the day is a wonder to us humans, who seek out the daylight.

Until quite recently, humans filled the vacuum of their knowledge with superstition and myths. This led people over the ages to be afraid of bats, to try to eliminate them from areas, and to invent stories about them that only increased the fear.

In today's world, the relationship between people and bats is changing. Scientists are learning about the vital role bats play in our environment in dispersing seeds, pollinating plants, and eating insects. People are attracting them to their yards with bat houses and chemical-free landscaping so they can enjoy watching them as well as have a natural control over many night-flying insects. And large organizations, such as the Organization for Bat Conservation, are promoting the understanding and conservation of bats through education, the media, and research. It is a great turn of events.

This is the first field guide to all of the bats of the United States and Canada that has ever been published. We have asked Rob Mies and Kim Williams to write it because they are at the forefront of the movement to appreciate and protect these animals. As founders and directors of the Organization for Bat Conservation, they bring a wealth of knowledge and enthusiasm to the task.

Of the 4,000 species of mammals in the world, nearly 1,000 of them are bats. There are 45 species of bats in the United States and Canada, and many of them are endangered. We hope this authoritative yet user-friendly guide will help you understand, identify, and appreciate the bats that may live near you, and that this in turn will lead you to take more active steps toward conserving these wonderful little mammals.

Yours in nature,
Don and Lillian Stokes

An Introduction to Bats

MAMMALS ON THE WING

Bats are the only mammals in the world that can fly. Because of this they are in their own order, called Chiroptera. Chiroptera is a Latin word that in English literally means "hand-wing." If you look closely at a bat's wing, you will notice that it has four fingers and a thumb (see the illustration on the last page of this book). The fingers are connected by a thin membrane of skin, making a wing. Aside from this, the biggest structural difference between a human's hand and a bat's wing is that the bat's fingers are greatly elongated.

Bats are remarkably agile fliers. If you watch a bat in your yard catching insects on the wing, you're sure to be amazed at the quick turns and its ability to catch a tiny insect in midair. Different species of bats have different wing characteristics, and these determine flight capabilities. A bat with short stout wings may be able to hover quite well and glean insects off leaves or the ground, whereas a bat with long narrow wings can probably fly quickly for long distances and catch insects easily in midair.

REPRODUCTION

Because bats are mammals, they breed and produce babies just as all mammals do. A male fertilizes the female internally, she becomes pregnant, and the young are born helpless and usually furless. Most bats must be at least 1 year old before mating. For many species of bats, fertilization may be delayed. For example, if a Big Brown Bat mates in fall, the egg and sperm are stored in the female's uterus until spring, when fertilization takes place.

Most species of bats in the United States and Canada have a 30- to 60-day gestation period. Most commonly a single pup is born. However, a few bat species may have twins, triplets, or even quadruplets. Mother bats nurse their young with milk until they are able to fly by themselves. Until this point she will leave them at a roost as she begins her nightly feeding. When her pups are very young, she will return often to check on them and nurse them. At roughly 3–5 weeks of age, most U.S. and Canadian bats will take their first flight. Presumably their mother takes them foraging with her to show them the ropes. Young will still be nursed until they are old enough to forage completely on their own.

ROOSTING

Bats are very selective homeowners and most have specific requirements for their homes. Once a bat or colony of bats picks a roost, they are often very loyal to it, returning night after night and year after year.

Most bats use cracks and crevices as roosting sites; examples include caves, mines, rocks, tree hollows, and under the loose bark of trees. Because these roost sites are generally in short supply, most crevice-dwelling bats are colonial; that is, they roost in groups. Roosting in large groups also makes it easier for the bats to stay warm.

There are also species of bats that use live foliage as roosts. These roosts are naturally less permanent than caves or shingles, and the bats that use them show much less fidelity to their homes. In addition, roosts in live trees are much more numerous; therefore many of these bats are solitary, choosing to live their lives alone.

For many species in temperate climates, winter roosts are different from summer roosts. Some bats choose these winter homes because they offer more suitable conditions for hibernation, and other bats seek out new homes in areas that offer their preferred foods throughout the winter. (See the section on hibernation and migration, page 10.)

Although bat species may differ greatly in the roost sites they use, they do share one attribute: all bats roost upside down. Why they do this remains a mystery to scientists. One theory is that in order to fly bats needed, evolutionarily speaking, to lighten their body weight load. They did this by reducing the density of their pelvic girdle, which made it impossible for them to stand on their hind legs. Another theory suggests that as bats began gliding from one tree to another, they simply hung under the branches of trees and let go when they wanted to take flight.

FEEDING

Worldwide, bats have a wide range of food preferences, feeding on fruit, nectar, insects, small vertebrates, blood, and even fish. In the United States and Canada, however, most species are

WHY ARE BATS IMPORTANT?

Bats are extremely important for humans and for the environment. Their importance ranges from dispersing seeds in the tropical rain forests to consuming millions of insect pests annually throughout the world.

Fruit bats, which are found mainly in the tropics, disperse seeds that help the rain forests grow, especially after widespread deforestation. The fruit bats eat the ripe fruit that grows on the trees, then spit or defecate the seeds out of their bodies. These seeds may grow into new trees.

Many types of bats drink nectar from flowers and at the same time pollinate the flowers they drink from, just as hummingbirds and butterflies do. As a matter of fact, the trees that bear many of the fruits we enjoy rely on bats for pollination.

Most bats in the United States and Canada feed on insects, not only the insects that bother us in our backyards, but also many that are detrimental to farmers' crops. This lessens the need for pesticides, which are unhealthy for us and cost farmers millions of dollars each year. One colony of Big Brown Bats can consume thousands of pounds of crop-damaging pests in a year. An individual bat can eat up to 1,200 mosquito-sized insects each hour! They eat up to their full body weight in insects each night. To put that in perspective, an adult human would need to eat roughly 80 large pizzas to eat what a bat eats each night.

insect eaters, with three species feeding on flowers and one on fruit.

Insectivorous bats have many options for finding food. Many fly over water, around light, or in open areas, like backyards where insects congregate. These bats use their echolocation ability (see page 19) to find prey. Once they have found the insect, they may either grab it with their mouth or scoop it up with a wing or tail membrane and transfer the food into their mouth to eat it.

Nectar-feeding and fruit-eating bats in North America do possess echolocation abilities. However, they rely more on sight, smell, and probably a good memorization of fruiting trees and flowering plants in their home range to find food.

Feeding behavior, and as a matter of fact nearly all bat behavior, takes place almost entirely at night. Being a nocturnal animal has many advantages for bats. First of all, being nocturnal ensures that bats will not be heavily preyed on. Second, it gives bats access to many more food sources; there are very few birds that eat insects, fruit, or flowers at night. And last, temperature is lowest and humidity highest at night, which helps bats avoid excess heat and water loss.

HIBERNATION AND MIGRATION

In temperate areas, where temperatures drop in the winter, bats hibernate beginning in fall because insects are not available. They wake in the spring when insect populations are again abundant. Hibernation is a state of deep sleep. During hibernation bats exist on the food that they have stored in their bodies as fat. They also lower their body temperature from roughly 100 degrees Fahrenheit to the temperature of their surroundings — usually between 40 and 50 degrees Fahrenheit.

Most bats in the United States use caves or mines as hibernating sites. These bats are very loyal to their hibernacula, returning to the same cave or mine each year, many times choosing to roost in exactly the same spot. Tree-roosting bats, for example red bats and Hoary Bats, do not use caves or mines to hibernate in. Instead, they usually choose hollow trees. These bats may not be as loyal to their hibernating sites.

Some bats may migrate every year, either to go where food is more plentiful or to hibernate. Many species fly long or even short distances from summer roosts to more favorable locales for their feeding activities. For example, nectar-feeding bats follow flower blooms, therefore insuring a food supply all year long. Other species migrate to favorable hibernating sites. Some bats fly long distances; for example, Mexican Free-tailed Bats from Texas may fly to Mexico for the winter. Others fly short distances; a Big Brown Bat in a northern state, for example, may fly in fall from one building across town to another building, where it will hibernate until spring.

How to Use This Guide

This *Stokes Beginner's Guide to Bats* includes the 45 species of bats found in the United States and Canada. Below is a description of the main parts of the guide and how to use them.

Please be aware that much information on bats is simply still not known. Therefore some sections have less detail than others; this is because the research is not available.

INDEX TO BATS

Inside the back cover is an alphabetical index of bat species. This helps you look up a species you are already familiar with, but about which you want more information. The bats are arranged in alphabetical order by their common names.

COLOR TAB INDEX

The Color Tab Index to Bat Genera on the first page of this book makes it easy to find the various family groupings of bats described in the Identification Pages. The brief description of each color-coded group may be all you need to lead you to the species identification account that matches the bat you have seen. Simply turn to the section of the book with the matching color tab and look through the species descriptions until you find your bat. With practice, you will soon become familiar with the bats most often seen in your area.

STATE LISTS

If you need more information than the Color Tab Index provides, the State Lists on pages 134–159 will help you quickly look up the specific bats found in each state of the United States and every province and territory of Canada. Once you find your state or province (they are listed alphabetically), you can see which bats have been recorded there and whether they are common or uncommon in that area. To check an identification, simply turn to the page indicated to read all about the species you are considering (your best bet is to check the common bats first).

IDENTIFICATION PAGES

Each species identification account starts with the bat's common name in large letters, followed by its scientific name in smaller italic letters. Bats of the same genus are grouped together and share the same color tab color (for example, all bats in the genus *Myotis* have a medium blue color tab). Some of the color tab sections contain more than one genus. Each species account includes the following further information:

I.D. — This section points out the main features of the bat that distinguish it from other species. Its description includes its length (L) from head to end of tail, wingspan (WS) from wing tip to wing tip, and weight (WT). It also tells you the bat's color and, if known, its lifespan.

In Flight — Describes the bat's typical way of flying. Many species of bats have distinguishing flight characteristics that will help you in identifying them.

Feeding — This section tells you when you are most likely to see the bat emerging from its day roost. It also indicates the most common areas where you might find the bat foraging for food.

Foods — Here you will find the foods — insects, fruits, nectar, for example — most commonly eaten by the species.

Echolocation Frequency — Different species of bats echolocate at different frequencies. This section shows the minimum and maximum frequencies, measured in kilohertz (kHz). The measures were derived from edited

Found Roosting

 In colonies, or groups

 In dead or dying trees (under loose bark, in cracks, crevices, old woodpecker holes, etc.)

 Individually

 In live trees (under leaves, in Spanish moss, hanging from branches, etc.)

 In caves, mines, or shafts

 In human-made structures (behind shutters, under shingles, in attics, in bat houses, under bridges, etc.)

 In rock crevices or cliffs

 Information unknown

Found Flying

 Around desert areas

 Over water

 Around flowers

 In open areas (fields, airports, golf courses, pastures, etc.)

 Around forested areas

 Along cliffs or rocky slopes

 Around edge habitat

 Around human habitation (backyards, city streets, streetlights, etc.)

Information unknown

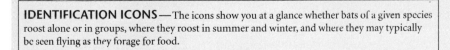

IDENTIFICATION ICONS—The icons show you at a glance whether bats of a given species roost alone or in groups, where they roost in summer and winter, and where they may typically be seen flying as they forage for food.

files using Analook (a computer program that analyzes bat echolocation calls from the Anabat detector) by Michael J. O'Farrell. Although echolocation frequency alone is not a foolproof means of identification, bat detectors are a useful and enjoyable way to monitor bat activity. (See pages 19–21 for a discussion of echolocation and bat detectors.)

Roosting — In general, bats roost and sleep during the day. Some bats roost alone, while others prefer to roost in small or large groups. Besides indicating roosting behavior, this section also describes the most common places the bat chooses as its roost during the summer months.

Migration/Hibernation — Here you will learn whether the bat migrates and/or hibernates or does neither.

This section also tells you where the bat chooses to roost during the winter months, if known.

Breeding — Tells you whether the bat forms maternity colonies or not, the number of offspring it has each year, when the young are born, and any other pertinent information.

Tip — Gives you information on how common the species is throughout the United States and Canada.

Range Map — The map gives you a general idea of where the bat lives. Winter and summer records of bats are poorly known; therefore the map encompasses its entire year-round range in the United States and Canada.

Because of the incomplete nature of current research on North American bat species, we have had to make some compromises in preparing the maps. When verified records show where the bat has been found, only that portion of the state is colored in; if the bat is verified as being present but there is no record of locale (or if the bat is found throughout the state), the entire state is colored. Similarly, if no research has been done on a species in a particular state, or if there are no documented sightings, then that state is left blank, even though the bat may in fact be present there. The hatching on some maps shows where a species has been recorded as an accidental; that is, it has been seen only rarely and its normal range is elsewhere.

DON'T BE AFRAID OF BATS

Throughout the centuries, people in many different cultures have been afraid of bats. They are often associated with Halloween and Count Dracula. However, bats are not the frightening creatures Hollywood has portrayed them to be. Here are some common myths, and truths, about bats.

MYTH: "Bats have rabies."

FACT: Like all mammals, bats can contract rabies. However, very few bats do. When bat specimens are sent to state laboratories for testing, the number of bats that test positive for rabies is usually around 5–10 percent. Even this number is misleading, because only bats suspected of having rabies are sent in to be tested. Research indicates that the actual incidence of rabies in bat populations is less than 0.5 percent in most areas.

Your chance of coming in contact with a rabid bat is very rare. In fact, you have a higher chance of winning your state lottery than of being bitten by a bat with rabies. Even though coming in contact with a rabid bat may be uncommon, it is wise to remember never to touch any wild animal.

MYTH: "Bats are blind and will get caught or nest in your hair."

FACT: First, bats are not blind. As a matter of fact, some bats have eyesight three times better than a human's. Second, a bat's echolocation ability is very precise, able to detect an object as fine as a single strand of human hair even in total darkness. Chances are very slight that a bat, with such accurate echolocation capabilities, would make a mistake and get tangled in your hair. Finally, bats do not make nests, they roost. Therefore a bat will not nest — in human hair or in anything else.

MYTH: "Bats attack humans."

FACT: Bats are very small and gentle animals; they will not attack people. Remember, we are very large to a bat, and they are afraid of us.

MYTH: "Vampire bats live in the United States and drink human blood."

FACT: Vampire bats live in Central and South America and in southern Mexico. They rarely drink the blood

A Common Vampire Bat. Vampire bats live only in southern Mexico and in Central and South America.

of humans, preferring the blood of chickens, cows, and pigs.

MYTH: "Bat droppings are dangerous."

FACT: Many people have heard of histoplasmosis, a fungal infection you could possibly get from bat or bird droppings. In order for the histoplasmosis fungus to develop, droppings must sit in a moist, humid place for a period of time. If the droppings are disturbed after the fungus grows on them, the spores from the fungus are released. A person who breathes in these spores may get histoplasmosis, which exhibits itself with flulike symptoms. It can be dangerous to people with asthma or other lung problems. To avoid histoplasmosis, remember to wear a face mask or respirator if you are in an enclosed area with animal droppings, to prevent inhaling the fungal spores.

MYTH: "A bat in your house is purposely trying to hit you as it flies around."

FACT: When bats accidentally find their way into a home, they are scared and try hard to find a way out. They are not, however, purposely trying to swoop at or attack you. In this situation, the person is usually standing in the center of the room, panicked, waving his arms and yelling. The terrified bat, meanwhile, is searching for an opening to escape. In order to fly around a room, the bat must fly a figure eight to gain the clearance to make the turns. It swoops down at the center of the room to give itself enough flight area to make the turn at the wall. If you just open a door or window and quietly sit down, the bat will probably find its way out.

BATS IN AND AROUND YOUR HOUSE

Bats Living in Your House — A bat's biggest problem is the shortage of naturally occurring places to live. Many species would live in a dead or dying tree, under the loose bark or in the hollows. But because many of these roosts are gone, a few bat species occasionally take up residence in human dwellings. In contrast to the old belief that the best way to get rid of bats is to kill them, today it is recommended instead to perform a humane exclusion — that is, to prevent the bats from returning to your house after they have left — and to provide alternative housing.

To do an exclusion, first locate the exit/entrance of the bats by watching the house from sunset till about 45 minutes after sunset; this is the period when they will be leaving to forage for food. Next, place a bat house near the area on a pole in the open, so the bats can find it when excluded from your house. Finally, construct a one-way device that allows the bats to exit but not return, and place it over the entrance.

This can be done in daylight hours by hanging a sort of curtain in front of the exit area (or areas) using a sheet of polypropylene netting (⅛- or ¼-inch mesh) or heavy-duty clear plastic. Attach the netting to your house at the top and sides — duct tape works fine (see photo, page 18). If necessary, place a strip of wood an inch or so deep above the exit hole, to help hold the netting out away from the house and avoid obstructing the exit. The netting or plastic sheet should extend several inches to either side of the exit area and hang a foot or two below it. The bats will be able to leave the building in the evening via the loose, open bottom of the curtain, but will not be able to find their way back in when they return.

Only do this in early spring, before pups are born, or in late summer, after young can fly. Leave the device up for at least a week to allow all the bats to leave. Once they are gone, you can

Because many natural bat roosts are destroyed each year, some bats may take up residence in homes.

make permanent repairs to the entrance/exit holes. The worst thing to do is trap a colony of bats in your house; not only is this inhumane, but they may find a way into your living quarters if their exit has been sealed.

Bats Flying Around in Your House —
If you ever have a bat flying around in your house, don't panic. If there is no

An exclusion — a one-way device that allows bats to exit a building but not reenter — is a simple and effective way to remove bats without harming them.

possibility that the bat has bitten anyone, simply open a door or window and let the bat fly out. You can also capture the bat using a butterfly net, box, thick towel, or heavy leather work gloves and release it, unharmed, outside. If you think that someone may have been bitten by the bat, safely capture the animal. Put the bat in a secure container, contact your local animal control, and instruct the officer to have it tested for rabies. If the test comes back positive (meaning the bat has the rabies virus), medical attention is needed. Contact your local health care provider.

Bats Found on the Ground — You should never handle wild animals, including bats, because you cannot tell if a mammal is ill. If you find a downed bat, the best thing to do is contact a local rehabilitator who has been trained to deal with bats. See www.batconservation.org under Bats and Humans for a listing of rehabilitators nationwide, or contact your local animal control agency. If a bat must be handled, take necessary precautions, using heavy leather work gloves or layers of toweling to scoop the bat up and move it to a safe place (e.g., in the woods, away from people).

Bats as Pets — Bats are wild animals and should always be left in the wild. It is illegal in most states to have a bat as a pet and a federal offense to possess a threatened or endangered bat species.

ECHOLOCATION

In the late 1700s, Italian scientist Lazarro Spallanzani theorized that bats could "see" with their ears. In the 1930s, Donald R. Griffin at Harvard University used a sensitive microphone to prove that bats produce high-frequency sounds whose echoes they use to avoid obstacles and identify food.

Echolocation is the term Griffin coined to describe this process. All U.S. and Canadian bats produce high-pitched, ultrasonic sounds with their vocal cords. These sounds strike objects in their path and bounce back to the bat's ears. The bat compares the outgoing sound to the incoming sound and forms an acoustic image of its environment. This is similar to the way sonar works in submarines. Other animals besides bats use echolocation, including dolphins, some species of whales, some species of shrews, and some species of birds.

Bats have a much more elaborately evolved echolocation ability than most other animals. They are able to detect large objects in their flight path and to identify and track small prey with remarkable accuracy.

The echolocation sounds are ultrasonic, which means they are usually at a higher frequency than humans can hear. As the bat flies around searching for food, it usually emits 10–50 calls per second. Once the bat detects an insect, the rate of the calls emitted will gradually increase to 200 or more per second. These shorter intervals give more detail about the insect's size, flight speed, and flight pattern. This high call rate is often called a "feeding buzz," because it represents an attack by the bat on its prey.

BAT DETECTORS

A bat's echolocation can be measured in kilohertz (kHz) using an electronic listening device called a bat detector. Most bats echolocate in the range of 20–120 kHz, with each species concentrating around one dominant frequency.

Bat detectors convert ultrasonic echolocation calls of bats into sounds audible to humans. Most bat detectors are considered broadband detectors, which means that they simultaneously detect a wide range of bat frequencies. Bat detectors can be used to detect the presence of bats, identify their foraging habitat, and occasionally determine the species. A tape recorder can be hooked up to the detector to record the converted echolocation for study at a later time.

This way of monitoring bats is becoming very useful for bat research. The major advantage is that the bats do not need to be seen or captured to document their presence in an area. With recent technological advances in computers, bat detectors are evolving with computer programs. Now some detectors (costing over $1,000) can be used by trained biologists to identify the species of bat with a good degree of accuracy. Basic broadband detectors are much less expensive ($80–$250) but of very good quality. Although you are unlikely to identify a particular

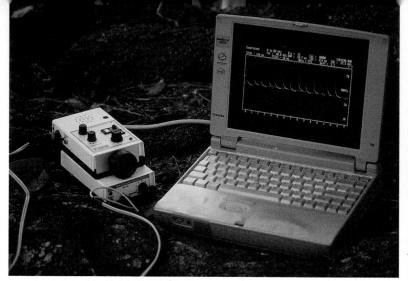

Bat detectors are one way scientists study bats.

species using one of these detectors, it is possible, with some practice, to identify the genus a bat is in.

The best place to listen for bats with the detector is around streetlights, near water, and around open areas where bats are foraging for food. When you are listening to the detector, other sounds can be heard. For instance, some insects give off ultrasonic sounds. On a bat detector, a bat's echolocation registers as a series of clicks or chirps. Listen for differences in the frequencies, which will indicate different bat genera. For instance, a Hoary Bat will be heard around 21–32 kHz and a Little Brown Bat around 38–62 kHz. You may also be able to tell when a bat is going after an insect. The "feeding buzz" will sound like fast, lower-frequency chirps on the detector.

Attracting Bats to Your Backyard

Bats are extremely beneficial animals. Many people, once they realize their value for insect control, like to encourage bats to live in their yards. Bats, like all animals, have three basic necessities for life: food, water, and shelter. Here are some ways to attract bats to your area.

- **Keep your backyard natural.** Keeping natural habitats — trees, woods, and especially dead snags — will provide homes for many bat species.
- **Put up bat houses.** Colony-roosting bats may use bat houses. As more natural habitat (dead trees) is destroyed, more bats will be looking for alternative places to roost. Bat houses, properly designed, simulate tree hollows and provide appropriate places for certain species of bats to live. See the next chapter for more information on bat houses.
- **Provide plants for roosting.** Solitary-roosting bats may use plants. Some species of bats, for example red bats, do not use cavities or bat houses. These bats roost in live trees, vines, and bushes. Certain types of plants may encourage a variety of bat species to roost in your backyard. Climbing vines like honeysuckle, ivy, white jasmine, and dog rose provide solitary species of bats places to roost.
- **Put in a pond or pool.** Installing a small pond or other water source may be a good way to attract bats.

Many people have seen bats skimming across their swimming pools at dusk, drinking the water.

- **Limit your use of pesticides.** Pesticides may kill insect pests in your yard, but they also harm bats. If you need to control your area for insects, try to use a natural pesticide like diatomaceous earth. Remember, the more bats that take up residence around you, the more insects they will eat, leaving fewer to bother you.

- **Turn on the lights.** Floodlights or porch lights attract insects. Attracting insects in turn attracts bats. You may then see bats foraging for insects around these lights.

- **Try night-blooming plants.** Certain types of plants that bloom at night attract insects. Flowers such as sweet rocket, evening primrose, nicotiana, and soapwort attract moths and other night-flying insects. Herbs such as chives, borage, mint, marjoram, and lemon balm also attract insects. This in turn attracts bats to feed on them.

- **Hang hummingbird feeders.** In the southwestern United States, hummingbird feeders may attract nectar-feeding bats. Some people report seeing hummingbirds at their feeders during the day and nectar-feeding bats enjoying the sugar water at night.

Certain night-blooming plants attract insects that may in turn attract bats to feed.

Bat Houses

Why would you want a bat house? Here are just a few of the reasons.

- **Insect control:** Bats can help control yard and garden insect pests.
- **A home of their own:** Bat houses give bats a place to roost other than *your* house. Some bats find their way into barns, garages, or older homes. Unfortunately, most people do not want bats living in their house, so they evict or kill them. If you want to exclude bats from your home or garage (see page 17 for details), putting a bat house nearby might give them a suitable alternative.
- **Replacing lost habitat:** Many species of bats in the United States and Canada would normally roost under the loose and peeling bark of dead or dying trees. Much of this naturally occurring habitat is gone, due to an ever-increasing human population and need for land. Therefore many bat populations are declining and need our help. One way you can help is to place a bat house on your property to provide roosting habitat for these displaced bats.
- **It's fun!** If you attract a population, you'll be able to watch the bats flutter as they look for insects, much the way you watch birds at your backyard feeder.

IMPORTANT FEATURES TO CONSIDER

Bat houses were first built in the 1950s. These early houses had dimensions similar to those of a birdhouse and were rarely used. Unfortunately some manufacturers still produce the 1950s-style houses. These have only a 10–30 percent chance of occupancy.

Starting in the 1990s, bat conservationists began testing many designs and features of bat houses to increase the occupancy rate. The first step in attracting bats to your bat house is to make sure it is an up-to-date model. The following are the main features you should keep in mind when building or purchasing a bat house:

- Weather-resistant wood is preferred (cedar or exterior-grade plywood).
- Screws are better than nails. Make sure they are galvanized.
- Caulk throughout the upper part of the house will keep the bats warm and dry.
- The inside of the house should have plastic mesh, grooves, or a very rough surface so the bats can hang easily.
- The house should be at least 24 inches tall and 14 inches wide, to allow the bats room to move around. Houses can be one or more chambers deep.
- The entrance slot to each chamber should be ¾–1 inch deep (no deeper!), to protect the bats from predators, such as blue jays, raccoons, or snakes.
- The house should have a landing area that extends 4–6 inches below the entrance, to allow the bats an easy way to get in and out.
- The house should have various temperature zones. A ceiling at the top of the house and a ventilation slot about a third of the way from the bottom will give the bats warmer and cooler areas to move to.

LOCATION, LOCATION, LOCATION

The second step in attracting bats to your bat house is to make sure you place it in the correct location. Houses mounted on poles and sides of buildings work much better than those placed on trees. Trees usually block the bats' view of the house. In addition, bats prefer warm roosts, and tree-mounted houses tend to be cooler.

Research has shown, by the way, that placing three or more houses in one area significantly increases your chance of getting bats into at least one of them.

Here are some points to keep in mind when choosing a site for your bat house:

- Bat houses should be mounted at least 12 feet off the ground, the higher the better. Bats like to be up high, away from predators. Also, in order to take flight, they usually need to free-fall several feet when leaving the house at night.
- Bats need an open area around the entrance to give them room to fly and to swoop into and out of the house. Make sure that there is at least 15 to 20 feet of clearance in front of the house.
- Bats like a warm place to raise their young. Bat houses should face south or southeast to take advantage of direct sunlight.

Bat houses, properly made and mounted, are a great way to help bats. This is a three-chambered bat house. Although from this angle the opening looks larger, the entrances to the chambers are the recommended ¾ inch deep.

In northeastern, northwestern, and midwestern states and in Canada, bat houses need to receive at least 6–8 hours of direct sunlight daily. We recommend that houses in those locales be painted black to absorb the sun's heat. Use a nontoxic, exterior latex paint or nontoxic stain, on the outside only.

In southeastern and southwestern states, some bat species like the house very hot and others like it cooler. Houses can be placed in partial sun (at least 4 hours a day) or in full sun. The bat house color can be dark or left natural. In extreme southern portions of the United States, houses can be painted white to reflect the sun and keep the bats cooler.

HOW DO YOU KNOW IF THE HOUSE IS OCCUPIED?

Bats usually start occupying a bat house in early spring after the hibernation season is

Big Brown Bats roosting in a multichambered bat house.

BUILDING YOUR OWN BAT HOUSE

In our work with the Organization for Bat Conservation we have been able to design houses that receive a very high occupancy rate! The simple design provided here has proven to work time after time in a variety of habitats. (If you have questions when building your bat house, or would like to purchase a bat house, check out OBC's Web site at www.batconservation.org.)

Bat houses should be made of exterior plywood or cedar (rough on the inside). The inside should have grooves at least every ¼ inch or be lined with ⅛-inch polyethylene plastic mesh (it's also called gutter guard, and is found at most large hardware stores) attached all the way up the front and back. Staple the mesh so it lies flat.

The house should be at least 24

over. Bats in some parts of the southern United States may use bat houses year-round. The easiest way to see if bats have moved in is to check with a strong flashlight in the morning, when they are least active. You can also place a piece of cardboard or screen under the house and look each day for guano (droppings). Another way is to watch the house around sunset to see them exit.

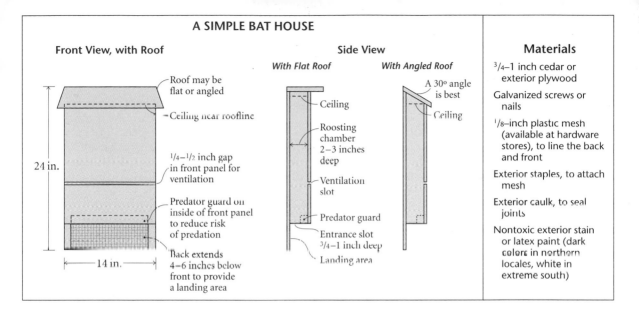

A SIMPLE BAT HOUSE

Front View, with Roof

Roof may be flat or angled

Ceiling near roofline

24 in.

1/4–1/2 inch gap in front panel for ventilation

Predator guard on inside of front panel to reduce risk of predation

Back extends 4–6 inches below front to provide a landing area

14 in.

Side View

With Flat Roof

Ceiling

Roosting chamber 2–3 inches deep

Ventilation slot

Predator guard

Entrance slot 3/4–1 inch deep

Landing area

With Angled Roof

A 30° angle is best

Ceiling

Materials

3/4–1 inch cedar or exterior plywood

Galvanized screws or nails

1/8-inch plastic mesh (available at hardware stores), to line the back and front

Exterior staples, to attach mesh

Exterior caulk, to seal joints

Nontoxic exterior stain or latex paint (dark colors in northern locales, white in extreme south)

---

inches tall, 14 inches wide, and 2–3 inches deep. It can be made with multiple chambers, although the design shown has just one.

The entrance slot, at the bottom of the house, should be about ¾–1 inch deep. Any larger will allow predators to enter. At OBC, we recommend

using 2-inch sides, then attaching a 1-inch-thick strip of wood to the lower front panel to create this small opening.

Caulking the house will keep the bats warm and dry, and putting it together with galvanized screws will help prolong the life of the house. A ceiling at the top of the house, just below the roof, and a ¼-inch ventilation space about 6 inches from the bottom opening, will create much-needed temperature variation.

The roof may be angled or flat. A roof at a 30-degree angle works best because it allows rain to properly run off the house.

Paint the outside of the house with two or three coats of exterior latex paint or nontoxic stain.

Bat Conservation

Until recently, there has been little known about bats and their benefits. Because of this, many bat species have become endangered and some have even disappeared entirely. Bats are one of our most essential allies. Worldwide, they are the major predators of nighttime insects, and many are key figures in plant pollination and seed dispersal.

Of the 45 species of bats found in the United States, 6 are currently considered endangered by the U.S. Fish and Wildlife Service. Another 20 species are considered of special concern and may be listed as threatened or endangered in the near future. Bat populations in general are declining. There are three major reasons:

- **Habitat destruction:** With an ever-increasing human population and our need for raw materials and space, crucial bat habitat is destroyed daily. This includes summer and winter roosting areas.
- **Fear of bats:** Many people grow up thinking bats will attack, have rabies, or will fly into their hair. Because of such beliefs, people have destroyed bats, sometimes in great numbers.
- **Chemical poisoning:** Pesticides sprayed on or near bats can be toxic, causing diminished reproductive efficiency or even death.

For these reasons, protection of all bat species and their habitat is necessary. There are a variety of conservation projects currently under way in the United States.

RESEARCH

Sampling for the presence, abundance, and species diversity of bats is being conducted in most states.

The most common way to study bats is to capture them in mist nets. Mist nets are large (10–60 feet long) nets made of very fine nylon thread. The nets are strung over ponds, streams, rivers, or dry land to capture bats as they fly at night foraging for insects.

Once captured, the bats are identified, sexed, aged, and weighed, and sometimes fitted with a tag or temporary radio transmitter. This transmitter allows researchers to track individual bats and determine foraging habitat, roosting preferences, and colony size. Other methods of

Authors Kim Williams and Rob Mies carefully remove a bat caught in a mist net. Once it has been identified, it will be released unharmed.

sampling include using night-vision scopes and bat detectors.

PROTECTING CAVES AND MINES

Bats often occupy mines and caves to raise their young in the summer or to hibernate in the winter. The largest numbers of bats in the United States are found in southern caves and mines, including the largest bat colony in the world (Bracken Cave, in Texas). Many of these caves and mines have been fitted with "bat friendly" gates that allow bats to enter and exit but keep people at a distance.

EDUCATION

Many organizations and agencies have recently instituted educational programs and projects as another conservation effort. Some zoos now have live bat exhibits that teach their visitors about the importance of bats and the need to protect them from extinction (see pages 32–35 for a list of places to view bats). Several nonprofit conservation organizations and governmental agencies provide educational material about bats to interested parties. The number of excellent bat documentaries shown on TV has risen in the last decade, and many cartoon movies are portraying

THINGS YOU CAN DO TO HELP BATS

- Support legislation that protects natural habitat and wildlife.
- Educate yourself through books, Web sites, videos, and other educational material.
- Teach others who are afraid of bats.
- Write letters to the editors of magazines and newspapers that print inaccurate information about bats.
- Join a conservation organization that works to protect bats.
- Put up a bat house in your backyard to give bats a place to live.
- Visit and support zoos that have bats on display or places that bats live naturally.
- Become a bat biologist or volunteer in research work.
- Support sustainable harvesting of trees, and plant more trees.
- Keep your property as natural as possible, including using few pesticides or none at all.
- Recycle, reuse, and reduce the things you buy and throw away.

bats in a kinder light. Educational books, videos, and Web sites have been produced that allow everyone a chance to learn the real facts about bats (see page 36 for some suggestions on where to learn more).

THE ORGANIZATION FOR BAT CONSERVATION

We founded the Organization for Bat Conservation in 1990 to help protect bats and their habitats through education, conservation, and ecological research. The OBC works on both a national and grassroots level. It presents more than a thousand educational programs each year for schools, nature centers, and zoos in the United States and Canada and has thousands of members around the world who share a concern for bats, the environment, and education. We invite you to find

Education is an important part of bat conservation. Here Rob Mies helps the New York City Parks Department learn more about these fascinating mammals.

out more about memberships, programs, or bat houses by contacting the Organization for Bat Conservation at 800-276-7074 or by checking out our Web site at www.batconservation.org.

Where to See Bats

MIDWEST

Brookfield Zoo
3300 Golf Rd.
Brookfield, IL 60513
Phone: 708-485-0263
Fax: 708-485-3532

Open 10 A.M.–5 P.M. year-round. **Bats to View:** Jamaican Fruit Bats and Vampire Bats on display. Rodriguez Fruit Bats are part of a walk-through exhibit.

Fort Wayne Children's Zoo
3411 Sherman Blvd.
Fort Wayne, IN 46808
Phone: 219-427-6800
Fax: 219-427-6820
Web site: www.kidszoo.com

Open Apr.–Oct. **Bats to View:** Walk-through rain forest exhibit has Indian Flying Foxes. Jamaican Fruit-eating Bats in a flight display.

Millie Hill Mine
On Millie Hill by Chapin Pit
Iron Mountain, MI
Contact: Tourism Association
Phone: 800-236-2447
Fax: 906-774-7739
Web site: www.ironmtntourism.org

Open year-round. **Bats to View:** Thousands of Little Brown, Big Brown, Eastern Pipistrelle, and Northern Long-eared Bats hibernate in this iron ore mine. The mine has a steel grate that keeps people out. Bats are best viewed in the spring and fall.

Milwaukee County Zoo
10001 W. Bluemound Rd.
Milwaukee, WI 53226
Phone: 414-771-3040
Fax: 414-256-5410

Open year-round. **Bats to View:** Vampire Bats in a nocturnal setting. Straw-colored Fruit Bats, Egyptian Fruit Bats, and Indian Fruit Bats are also at this zoo.

Organization for Bat Conservation
1553 Haslett Rd.
Haslett, MI 48840
Phone: 800-276-7074
Fax: 517-339-5618
Email: obcbats@aol.com
Web site: www.batconservation.org

Tours to members Mon.–Fri. 9 A.M.–5 P.M. **Bats to View:** A variety of native nonreleasable bats, including Big Brown Bats, Pallid Bats, Evening Bats, Brazilian Free-tailed Bats, and others. Exotic, captive-bred bats include Egyptian Fruit Bats, Straw-colored Fruit Bats, Greater Spear-nosed Bats, Mexican Long-tongued Bats, Vampire Bats, Jamaican Fruit-eating Bats, and the endangered Rodriguez Fruit Bat.

Wyandotte Caves
7315 S. Wyandotte Cave Rd.
Leavenworth, IN 47137
Contact: Bill Schulze
Phone: 812-738-2782
Fax: 812-738-8255
Web site: www.cccn.net

Bats to View: Endangered Indiana Bat, Little Brown Bat, Eastern Pipistrelle Bat, and Big Brown Bat.

SOUTHEAST

Gainesville Bat House
On Museum Rd. across from
   Lake Alice
University of Florida
Gainesville, FL
Contact Person: Ken Glover (UF)
Phone: 352-392-1904
Fax: 352-392-6367
Email: kglover@ehs.ufl.edu
Web site: www.ehs.ufl.edu

Open year-round. **Bats to View:** Large bat house (18 × 18 feet) built to relocate bats from other campus buildings now houses up to 70,000 bats. Just after sunset, Mexican Free-tailed Bats and Southeastern Bats can be seen exiting this manmade structure.

Lubee Foundation
1309 N.W. 192nd Ave.
Gainesville, FL 32609

Phone: 352-485-1250
Fax: 352-485-2656
Email: lubeebat@aol.com
Web site: www.lubee.com

Open by appointment. **Bats to View:** Lubee mainly focuses on Old World fruit bats. Twelve species of bats are housed at this captive breeding and research facility. Some of the bats include Rodriguez Fruit Bats, Indian Fruit Bats, and Pumila Fruit Bats.

Organization for Bat Conservation's Florida Bat Center
10941 Burnt Store Rd.
Punta Gorda, FL 33955
Phone: 941-637-6990
Email: flabats@aol.com
Web site: www.floridabats.org

Open 9 A.M.–5 P.M. by appointment. **Bats to View:** A variety of native non-releasable bats on exhibit, including

Brazilian Free-tailed Bats, yellow bats, Big Brown Bats, and Evening Bats. Straw-colored Fruit Bats, Egyptian Fruit Bats, and Mountain Fruit Bats are shown during their educational programs.

NORTHEAST

Bronx Zoo
100 85th and Southern Blvd.
Bronx, NY 10460
Phone: 718-220-5100
Web site: www.bronxzoo.com

Open year-round. **Bats to View:** "Jungleworld" has a variety of animals in a large free-flight building, includ-ing Rodriguez Fruit Bats. Jamaican and Short-tailed Fruit Bats are in the nocturnal house.

Canoe Creek State Park
RR2, Box 560
Hollidaysburg, PA 16648-9752

Phone: 814-695-6807
Fax: 814-696-6023
Web site: www.dcnr.state.pa.us

Open year-round. **Bats to View:** Little Brown Bats can be seen exiting by the thousands from an abandoned church. Bats can also be seen living in numerous bat houses and bat condos at the park.

Philadelphia Zoo
3400 W. Girard Ave.
Philadelphia, PA 19104
Phone: 215-243-1100
Fax: 215-243-0219
Web site: www.phillyzoo.org

Open 9:30 A.M.–5 P.M. year-round. **Bats to View:** Rodriguez Fruit Bats and Egyptian Fruit Bats can be seen in a big open flight cage lit by natural daylight. Vampire Bats can be seen in a cave display.

NORTHWEST

Oregon Zoo
4001 S.W. Canyon Rd.
Portland, OR 97221
Phone: 503-226-1561
Fax: 503-226-6836
Web site: www.zooregon.org

Open every day except Christmas; hours are roughly 9 A.M.–5 P.M. depending on season. **Bats to View:** Egyptian Fruit Bats, Rodriguez Fruit Bats, Straw-colored Fruit Bats, and Jamaican Fruit-eating Bats can be viewed in a horseshoe-shaped exhibit that simulates an African rain forest.

Woodland Park Zoo
5500 Phinney Ave. N.
Seattle, WA 98103
Phone: 206-684-4800
Fax: 206-684-4854
Web site: www.zoo.org

Open year-round. **Bats to View:** Vampire Bats, Straw-colored Fruit Bats, and Grey-headed Flying Foxes are on exhibit in the "Night Exhibit."

SOUTHWEST

Bandelier National Monument
HCR 1, Box 1, Suite 15
Los Alamos, NM 87544
Phone: 505-672-3861

Open year-round except Christmas and New Year's. **Bats to View:** Several thousand bats, mainly Brazilian Free-tailed Bats, can be viewed exiting a cave April through Sept.

Bat World
217 N. Oak Ave.
Mineral Wells, TX 76067
Phone: 940-325-3404
Web site: www.batworld.org

Carlsbad Caverns in New Mexico is a wonderful place to view bats.

Scheduled tours by appointment. **Bats to View:** Native nonreleasable bats can be seen in indoor cages. Most include Mexican Free-tailed Bats rescued in Texas. Egyptian and other fruit bats on display in flight cages.

Carlsbad Caverns National Park
3225 National Parks Hwy.
Carlsbad, NM 88220

Phone: 505-785-2232
Fax: 505-785-2302
Web site: www.nps.gov/cave

Open year-round, except Christmas. **Bats to View:** Hundreds of thousands of primarily Brazilian Free-tailed Bats emerge nightly from the caverns May through Oct. Amphitheater has an interpretive program nightly.

Congress Avenue Bridge
Congress Ave. between Riverside Dr.
and First St. (Cesar Chavez)
Austin, TX
For more information: 512-416-5700
ext. 3636

Open year-round. **Bats to View:**
Hundreds of thousands of Brazilian
Free-tailed Bats can be seen exiting from
under the bridge March through Oct.

Where to Learn More About Bats

BOOKS

Understanding Bats. Kim Williams
and Rob Mies. Marietta, Ohio: Bird
Watchers Digest Press, 1996.

**A Simple Guide to Bat House
Designs.** Organization for Bat
Conservation. Haslett, Mich.:
Organization for Bat Conserva-
tion, 1997.

VIDEO

Bats — The True Story. Haslett, Mich.:
Organization for Bat Conservation,
1998. Award-winning 42-minute
video provides great bat information
in an easy-to-understand context.
Wonderful footage of live bats.

ADOPT-A-BAT PROGRAMS

The Organization for Bat
Conservation offers a wonderful
Adopt-a-Bat Program. You can adopt a
bat of your choice to help with bat
conservation work. Sponsors receive a
poster, a letter of thanks from their bat,
and information about their bat with a
photo. Contact the OBC for details.

INTERNET

Basically Bats
www.lads.com/basicallybats

Bat Conservation International
www.batcons.org

Bat Conservation of Wisconsin
www.batcow.org

Bat Conservation Society of Canada
www.cancaver.ca/bats/canada.htm

Acknowledgments

Bat Conservation Trust
www.bats.org.uk

Bats Northwest
www.batsnorthwest.org

Bat World Sanctuary
www.batworld.org

Buzbee Bat House
www.batbox.org

Carlsbad Caverns Guadalupe Mountains Association
www.caverns.org

Lubee Foundation
www.lubee.com

Organization for Bat Conservation
www.batconservation.org

Organization for Bat Conservation's Florida Bat Center
www.floridabats.org

The authors would like to express their deep appreciation to the following people, whose expertise in reviewing, commenting, and adding technical information contributed greatly to this book:

Bill Gannon, Ph.D., University of New Mexico
Bill Korn, Ph.D., University of Florida
Allen Kurta, Ph.D., Eastern Michigan University
Cyndi and George Marks, OBC's Florida Bat Center
Philip Myers, Ph.D., University of Michigan–Ann Arbor
Michael J. O'Farrell, Ph.D., O'Farrell Biological Consulting
Scott C. Pedersen, Ph.D., South Dakota State University
Tim Strickler, Ph.D., Grand Valley State University, Michigan
Eugene H. Studier, Ph.D., University of Michigan–Flint
John O. Whitaker, Jr., Ph.D., Indiana State University

Thanks also to persons in each state and province who supplied information about species and conservation status.

Alaska: Doreen Parker McNeill, Alaska Department of Fish and Game
Alberta: Robert Barclay, University of Calgary
Arizona: Tim Snow, Arizona Game and Fish Department
Arkansas: Harry Harnish, park interpreter, Devil's Den State Park, and David Saugey

British Columbia: *Bats of British Columbia,* by David W. Nagorsen and R. Mark Brigham. Published by UBC Press in collaboration with the Royal British Columbia Museum.

California: Dr. Daniel F. Williams, California State University

Colorado: Michael Wunder, Colorado Natural Heritage Program

Connecticut: Jenny Dickson, Connecticut Department of Environmental Protection

Delaware: Christopher M. Heckscher, Delaware Natural Heritage Program

Florida: Cyndi Marks, Organization for Bat Conservation's Florida Bat Center

Georgia: Jim Ozier, Georgia Department of Natural Resources

Illinois: Joe Kath, Illinois Department of Natural Resources

Indiana: Scott Johnson, Indiana Department of Natural Resources

Iowa: Kim Bogenschutz, Iowa Department of Natural Resources

Kansas: Charlie Lee, Kansas State University

Kentucky: Traci Wethington, Kentucky Department of Fish and Wildlife Resources

Maine: Karen Morris, Maine Department of Inland Fisheries and Wildlife

Massachusetts: D. Scott Reynolds, Boston University

Michigan: Dr. Allen Kurta, Eastern Michigan University

Minnesota: Gerda Nordquist, Minnesota Department of Natural Resources

Mississippi: Richard Rummel, Mississippi Department of Wildlife, Fisheries, and Parks

Missouri: Missouri Department of Conservation

Montana: Amy Kuenzi, Department of Biology, Montana Tech

Nebraska: Patricia Freeman, Professor/Curator of Zoology, University of Nebraska State Museum, and Kenneth N. Geluso, Department of Biology, University of Nebraska at Omaha

Nevada: Michael O'Farrell, O'Farrell Biological Consulting

New Brunswick: Graham Forbes and Hugh Broders

New Hampshire: John Kanter, New Hampshire Fish and Game

New Jersey: Mike Valent, New Jersey Department of Environmental Protection, Division of Fish, Game, and Wildlife

New Mexico: David Roemer, biologist, Carlsbad Caverns National Park

New York: Alan Hicks, New York State Department of Environmental Conservation, Wildlife Diversity Section, Endangered Species Unit

North Carolina: Mary Kay Clark, North Carolina Museum of Natural Sciences

North Dakota: North Dakota Parks and Recreation

Ohio: Dave Swanson, Ohio Department of Natural Resources, Division of Wildlife

Oklahoma: Mark Howery, natural resource biologist

Ontario: Ojibway Nature Center

Oregon: Steve Langenstein, Bureau of Land Management

Pennsylvania: Penn State College of Agricultural Sciences Cooperative Extension

Rhode Island: Chris Paithel, Rhode Island Department of Environmental Management, Division of Fish, Wildlife, and Estuarine Resources

Saskatchewan: Mark Brigham, professor, Department of Biology, University of Regina

South Carolina: Mary Bunch, South Carolina Department of Natural Resources

Tennessee: Gary McCracken, University of Tennessee

Texas: *The Bats of Texas,* by David J. Schmidly, Texas A&M Press

Utah: George Oliver, Utah Natural Heritage Program, Division of Wildlife Resources

Vermont: Steve Parren, Vermont Fish and Wildlife Department, Non-game and Natural Heritage Program

Virginia: Carol Ann Curran, Virginia Beach Science Museum

Washington: Bats Northwest

West Virginia: Dr. Phillip Clem, University of Charleston

Wisconsin: Joseph Senulis, Wisconsin Department of Natural Resources

Wyoming: Martin Grenier, non-game mammal biologist, Wyoming Game and Fish

Special thanks to Ron and Debbie Williams for their help compiling information and data.

To help us compile information about bats, status, and ranges, we consulted the relevant species accounts in *Mammalian Species*, a publication of the American Society of Mammalogists (an index to this series is available at www.science.smith.edu/departments/biology/vhassen/msi_intro.html), as well as the following books and references:

Bats of America. Roger W. Barbour and Wayne H. Davis. Lexington: University Press of Kentucky, 1969.

Bats of the United States. Michael J. Harvey, J. Scott Altenbach, Troy L. Best. Little Rock: Arkansas Game and Fish Commission, 1999.

Mammals of the Eastern United States. John Whitaker, Jr., and

William J. Hamilton, Jr. Ithaca,
N.Y.: Cornell University Press, 1998.

Mammals of the Great Lakes Region.
Allen Kurta. Ann Arbor: University
of Michigan Press, 1995.

**The Smithsonian Book of North
American Mammals.** Don E.
Wilson and Sue Ruff, eds.
Washington, D.C.: Smithsonian
Institution, 1999.

Understanding Bats. Kimberly J.
Williams and Rob Mies. Marietta,
Ohio: Bird Watchers Digest Press,
1996.

Identification Pages

Ghost-faced Bat

Mormoops megalophylla

Roosting

in summer

in winter

Flying

I.D.: L: 3.1–3.9" WS: 14–16" WT: 0.5–0.6 oz. A large bat that varies in color from reddish to brown. Face has folds of skin that stretch across the chin from ear to ear, giving it a bizarre appearance.

In Flight: Flies strong and fast, relatively high above the ground, between foraging sites.

Feeding: Emerges after dusk. Found feeding in desert and riverine habitats in lowland areas. This bat has been seen foraging over standing water.

Foods: Eats mostly large moths but may eat other insects. In Mexico, local people use the Ghost-faced Bat's guano (feces) for fertilizer because, like the droppings of most other bats, it is so rich in nitrogen from the prey insects' exoskeletons.

Echolocation Frequency: 47–53 kHz.

Roosting: Roosts with other bats of the same species in caves, mines, and tunnels. It does not form clusters with others; instead, members of the colony roost singly, spread out about 6 inches from one another. Females with young roost separately from males and nonreproductive females. When sleeping during the day, it roosts with its back arched and head tucked up toward chest. Colony size may reach 500,000 individuals.

Migration/Hibernation: Roosts in caves and buildings during winter months but probably does not hibernate.

Breeding: Forms nursery colonies in deep warm areas of caves. One pup is born in late May or early June.

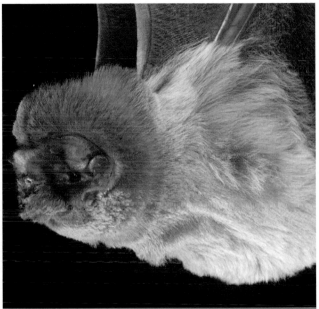

Tip Common winter resident in caves along southern edge of the Edwards Plateau, Tex. May also be regularly seen in the Big Bend area. Rare elsewhere.

California Leaf-nosed Bat

Macrotus californicus

Roosting

in summer

in winter

Flying

I.D.: L: 3.4–3.9" WS: 13–15" WT: 0.4–0.8 oz. A medium to large bat with long ears (longer than 1 inch) and a spear-shaped flap of skin that protrudes above the nose. It is a grayish color, with the fur nearest the skin almost white. It has large eyes. Can live 14 years or more.

In Flight: Very agile and maneuverable. When searching for food it flies slowly and quietly, although it can sometimes fly rapidly. Can be seen hovering in flight with its body vertical, ears standing up straight, head forward, and legs wide apart with tail membrane spread. The wingbeats are rapid and shallow while hovering.

Feeding: Emerges roughly 30 minutes to 1 hour after sunset, with small groups of bats leaving together over a 3-hour period. Many food items are nonflying; these prey are probably picked from the ground or from vegetation. Regularly seen foraging over the ground in desert areas and close to vegetation. Has a small foraging range; therefore usually seen searching for food less than 1 mile from its roost.

Foods: Food items include mostly moths, butterflies, and katydids, but this bat will also eat grasshoppers, cicadas, caterpillars, and beetles. Remains of dragonflies have been found beneath roosts, making these insects likely to have been eaten as well.

Echolocation Frequency: 56–72 kHz.

Roosting: Abandoned mine tunnels are its usual daytime roost. When roosting, it usually hangs by one foot, with the other leg dangling to the side or scratching and grooming its fur. Usually found in groups, but does not form tight clusters. When roosting, bats do not usually touch one another. Night roosts may be different from daytime roosts and include open buildings, cellars, porches, rocks, and mines.

Migration/Hibernation: These bats do not hibernate but congregate in warm mines (especially in Ariz. and Calif.) during the winter months. They are less active during this period of time.

Breeding: Male bats attract females by flapping their wings and calling to them (the call is audible to humans). Mating and fertilization occur during Sept., Oct., and Nov. Development of the fetus is delayed until spring. Gestation is roughly 9 months. Females congregate in maternity colonies of 100 to 200. Usually one pup is born, but sometimes twins, between mid-May and mid-July. Young males do not mate their first year.

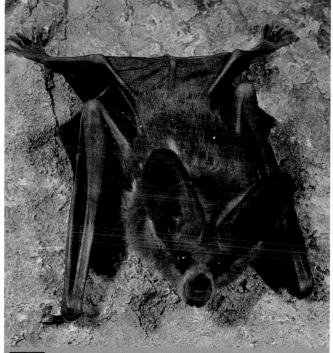

Tip Appears to be locally common in some areas, but rare throughout most of its range. More information is needed to determine the status of this bat.

Mexican Long-tongued Bat

Choeronycteris mexicana

Roosting

in summer

in winter

Flying

I.D.: L: 3.2–4.1" WS: 13–15" WT: 0.4–0.9 oz. A medium to large bat with a long and slender nose. Has a spear-shaped flap of skin protruding from the end of its nose. Has large eyes. Color ranges from gray to brownish. This bat has a short but conspicuous tail.

In Flight: Wings make a swishing sound. This bat may be able to hover in flight.

Feeding: Can be found feeding on flowers and insects in a variety of vegetative habitats, which include shrub to tropical forests and oak-conifer forests.

Foods: Feeds on nectar, pollen, fruits, and insects. These bats have been seen with heads and faces covered with pollen. They are effective and important pollinators of many flowers.

Echolocation Frequency: 41–74 kHz.

Roosting: Caves and abandoned mines are favored roosts, but these bats also can be found in buildings and culverts. When roosting, they choose dimly lit areas around the entrance of the roost. They often hang from one foot. When disturbed, these bats take flight quickly and usually fly into the daylight instead of deeper into the roost. They are colonial but do not cluster; instead they hang 1–2 inches away from one another.

Migration/Hibernation: Mexican Long-tongued Bats probably migrate seasonally to take advantage of food availability (for example, when flowers are in bloom). They probably do not hibernate. Researchers have reported bats leaving N.Mex. in Aug., leaving

Ariz. in Oct., and leaving Calif. in Dec. The bats probably migrate to Mexico.

Breeding: Usually one pup is born in June or July. Babies are born well furred and are roughly 30 percent of the mother's weight.

Tip Considered rare in the United States.

Lesser Long-nosed Bat

Leptonycteris curasoae

Roosting

in summer

in winter

Flying

I.D.: L: 3.0–3.4" WS: 14–16" WT: 0.5–0.9 oz. A medium to large bat with a long nose and a spear-shaped flap of skin at the end of its nose. This bat has no tail and has large eyes. Color is grayish to brownish toward skin and brownish at ends of hairs.

In Flight: Agile flier and able to fly nearly straight up in the air while maintaining a horizontal body position. Also able to hover. Flight is rapid and direct, not erratic as you might see with bats that consume insects. Strong wingbeats make a roaring sound that is very distinctive.

Feeding: Emerges about 1 hour after sunset and forages in desert-scrub areas where agaves, yuccas, saguaros, and organ pipe cacti grow. Bats can land on flowering stalks and insert their long noses into the flowers. They may also fly close to flowers when feeding. Some people report watching these bats feed from their hummingbird feeders during the night.

Foods: Eats nectar, pollen, and insects. Plants from which the bat drinks nectar include flowering stalks of the agave, saguaro, and organ pipe cacti. If the bats do not get all the nectar in one feeding they return again and again until the nectar is gone. After the flowering season is past, fruits are eaten. Several species of agaves depend on this bat for pollination.

Echolocation Frequency: 44–60 kHz.

Roosting: Found roosting in colonies, from a few bats to more than 10,000, in caves and mines at base of mountains. Hangs with feet close together and head downward. When disturbed, it raises its head to look at the intruder. Shy, and flies quickly when disturbed.

When bothered, it gives several strong wingbeats to bring its body to a horizontal position before releasing its feet from the roost and taking flight.

Migration/Hibernation: During Sept., this species of bat may leave the U.S. to winter in Mexico. It returns to U.S. in early May, following flower blooms.

Breeding: In early May, females form maternity colonies, which may number into the thousands. A single pup is born in mid- to late May and may be flying by late June. By late July, these large maternity colonies are already splitting up and females are then roosting in smaller numbers. The males stay in small groups while young are raised by the females.

Tip Endangered species and not often seen. However, in southernmost Ariz. and N.Mex. may be seen feeding at flowers.

Greater Long-nosed Bat

Leptonycteris nivalis

Roosting

in summer

in winter

Flying

I.D.: L: 3.0–3.5" WS: 16–17" WT: 0.6–1.1 oz. A large bat with a spear-shaped flap of skin on the end of its nose. Color is sooty brown, with the base of the hair white and the tips silver. This bat has no tail.

In Flight: A very strong and direct flier. When flying, makes a swishing sound with its wings. It is very agile and even able to fly straight up in the air while keeping its body in a horizontal position.

Feeding: Emerges late in the evening. Forages after dark in dry rocky canyons among agave and desert plants. Although this species may be found roosting in higher elevations, it usually moves to lower elevations to feed and forages on hillsides at these lower elevations. When feeding on agave, it lands on the stalk and crawls down to the flower. It then sticks its long tongue into the flower and laps the nectar.

Foods: Feeds on nectar and pollen from flowers like the century plant. It is a very important pollinator of many cacti, agaves, and other desert plants.

Echolocation Frequency: Unknown.

Roosting: Colony-roosting bat that lives deep in caves but has also been found in mines. Colonies in the southern extent of its range consist of fewer than 500 individuals, whereas in Tex. colonies can be larger than 10,000. It is said that a colony of these bats can be recognized by their musky smell, which is similar to that of the Brazilian Free-tailed Bat (p. 130).

Migration/Hibernation: In winter this bat apparently leaves the U.S. to winter in Mexico. It probably does not hibernate.

Breeding: Reproduction is poorly studied in this bat. One pup is born in April, May, or June. By July young are almost full-grown and are flying.

Tip Very rare in the U.S. Populations have been found in Brewster and Presidio Counties in Trans-Pecos, Tex., and Hidalgo County, N.Mex.

Jamaican Fruit-eating Bat

Artibeus jamaicensis

Roosting

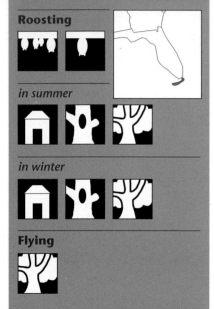

in summer

in winter

Flying

I.D.: L: 3" WS: 17–19" WT: 1.5–1.6 oz. A large bat with brownish fur. This is the only fruit-eating bat in the eastern U.S. It has a spear-shaped flap of skin on the end of its nose. Can live 7–10 years or more.

In Flight: A quick and powerful flier. Very agile when flying through trees. Is less active on bright nights with a full moon.

Feeding: Often picks fruit and carries it in its mouth to another location to eat. Forages in small groups.

Foods: Eats mostly fruits like green figs from the genus *Ficus*. Also enjoys mangoes, bananas, and avocados. It will also eat insects, pollen, and nectar.

Echolocation Frequency: 25–45 kHz when feeding.

Roosting: Roosts in buildings, hollow trees, and under tropical leaves. Roosts in harems of up to 25 females and 1 male. Males will also form bachelor groups, sometimes with nonreproductive females as well. Males may also be solitary.

Migration/Hibernation: Probably does not migrate or hibernate.

Breeding: Reproduces when fruit is most abundant. May reproduce twice a year. One pup, occasionally twins, born from March through April; a second birth may occur from Sept. to Dec. Females leave pups in roost as they forage at night.

Tip Rare in the U.S. Has only been found on Cudjoe and Ramrod Keys and Key West, Fla. It may occur on other keys as well, especially the southern keys.

Pallid Bat
Antrozous pallidus

Roosting

in summer

in winter

Flying

I.D.: L: 3.6–5.3" WS: 15–16" WT: 0.5–1.0 oz. A large bat with long ears and broad wings. Color on back is pale and yellowish, with brown- or gray-tipped hairs. Belly is almost white. Females are larger than males. Have lived up to 9 years in captivity; however, their lifespan in nature is unknown.

In Flight: Fly 2–3 feet above the ground as they head for water or feeding areas. Flight is noisy. The wingbeats are slow and fluttery. These bats may fly close to the ground when searching for food and may hover momentarily. Flight also may include gliding short distances.

Feeding: Emerges about 45 minutes after sunset. Feeds by taking prey off the ground. The Pallid Bat listens for its prey with its large ears. It then lands on the ground and crawls to catch its food. May also forage among trees, where it catches prey on the leaves. Very few of its prey are caught in the air.

Foods: Food items taken from the ground include scorpions, crickets, centipedes, beetles, and grasshoppers. Prey gleaned from leaves include cicadas, katydids, praying mantises, beetles, and moths. Pallid Bats may even eat lizards and rodents.

Echolocation Frequency: 28–49 kHz.

Roosting: Roosts in colonies of roughly 20 or more bats in buildings and rock crevices. May also roost in caves, mines, rock piles, and tree cavities. These bats tend to choose roosts where they can easily retreat into tight crevices when disturbed.

Tip A locally common species at lower elevations, especially in western United States

Males and females tend to roost separately during the maternity season. On warm days, the bats can be heard squabbling in their roosts. They have a characteristic odor, faintly skunklike, that can be smelled even from outside the roost.

Migration/Hibernation: Pallid Bats are thought to overwinter in the same vicinity in which they spend the summer. Although Pallid Bats do hibernate in some areas, in low to middle elevations they may remain active throughout winter. Summer and winter roosting sites (e.g., rock crevices and buildings) are also the same. They have been found hibernating singly or in pairs. Large hibernating groups are rare.

Breeding: Mating occurs from Oct. through Feb. Sperm is stored in the female's body for several months. Females become pregnant in spring. Gestation is roughly 9 weeks. Colonies of pregnant bats form in early April. Young are born in April, May, and June. Young females commonly give birth to one pup, whereas older females usually give birth to two. At birth, eyes are closed, pups are furless, and ears are folded against the head. Young can fly at 6 weeks of age.

Big Brown Bat

Eptesicus fuscus

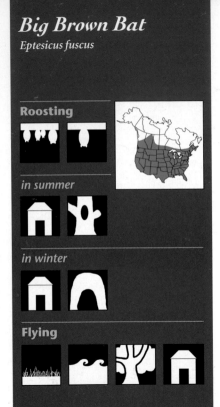

Roosting

in summer

in winter

Flying

I.D.: L: 3.4–5.4" WS: 13–16" WT: 0.4–0.8 oz. A medium to large brown bat that varies in color from light to dark brown. The wings, tail, ears, and nose lack fur and are dark brown to black. Females are larger than males. Can live at least 19 years.

In Flight: Flies a steady, nearly straight path at a height of 20–30 feet. Its large size and steady flight make it recognizable. Flight appears slow, probably due to its large size and slow wingbeats. Often flies the same course every evening and usually with other bats of the same species.

Feeding: Emerges about 20 minutes after sunset. May fly near roost at first, but then moves off to feeding grounds. Feeds on insects in cleared meadows, over water, among trees in pastures, along streets, in backyards, and even above traffic in the middle of the city.

Foods: Big Brown Bats eat beetles, ants, flies, leafhoppers, mayflies, stoneflies, and many other types of insects. They are very efficient feeders, able to consume up to their entire body weight in insects each night. Eat many agricultural pest insects, such as spotted cucumber beetles and scarab beetles.

Echolocation Frequency: 25–51 kHz.

Roosting: Daytime roosts are in dark places, usually in buildings or trees. In the East favored roosts include attics, barns, behind shutters, in bat houses, and under bridges. In the West hollow trees are commonly used. Night roosts include porches, garages, or barns with open doors. Throughout the spring and summer the females form maternity colonies of 20–500 bats, while the males are usually solitary.

Migration/Hibernation: Usually migrate only a short distance in Sept. to hibernate. Hibernating bats can be found in buildings, caves, mines, houses, or other shelters. In the East they are found hibernating mainly in buildings. Most of these bats are solitary. Females return to the same summer roost in March or April.

Breeding: Mates in fall, and fertilization is delayed until spring. After a 60-day gestation, females give birth to one or two pups in spring or early summer. Pups are born with eyes closed and are furless. Juveniles can fly and forage on their own after 3–4 weeks.

Tip Considered one of the most abundant and commonly seen bats in North America.

Spotted Bat
Euderma maculatum

Roosting

in summer

in winter

Flying

I.D.: L: 4.2–4.9" WS: 13–15" WT: 0.5–0.8 oz. A large bat with a very distinctive look. This bat cannot be confused with any other bat in North America. Its body is black with three large white patches of fur, one patch on each shoulder and one patch on the rump. Its ears are the largest of any North American bat's and are pinkish.

In Flight: Can fly high in the air, at elevations of 33–50 feet, usually above the trees. It has a direct and rapid flight pattern. It often hunts the same areas each night. It is very specific about the times it forages at each site, usually visiting areas near the same time each night. You can hear a Spotted Bat's echolocation call, which sounds like a sharp click, from as far away as 825 feet or more. It is a loud high-pitched sound that is clearly audible to our ears.

Feeding: Emerges from roosts 30–60 minutes after dusk. Returns to roost, after feeding, an hour before sunrise. It is a solitary hunter that feeds on insects in clearings. It may fly up to 6.2 miles between roosts and feeding grounds.

Foods: Food items include primarily moths. It has been speculated that this bat's low-frequency calls enable it to avoid being detected by species of moths that can detect the higher-frequency echolocation calls of most other bats.

Echolocation Frequency: 8–12 kHz.

Roosting: Spotted Bats roost in rock crevices on high cliffs. They can crawl very easily on both horizontal and vertical surfaces of these cliffs. The large ears are usually kept folded down along the back when roosting but may stand erect when the bat becomes alert or when it is preparing to take flight.

Migration/Hibernation: Little is known about the migration and hibernation habits of the Spotted Bat throughout its range. This bat is known to hibernate for short intervals, but it may remain active throughout the winter at low to middle elevations.

Breeding: Mating probably occurs in autumn, and sperm is stored in the female's body until spring. One pup is usually born in June, and newborn babies lack the beautiful black-and-white coloring that adults have. The newborn baby is roughly 20 percent of the mother's weight. The eyes are closed at birth, and the ears are large and floppy.

Tip Rarely seen in the United States.

Allen's Big-eared Bat

Idionycteris phyllotis

Roosting

in summer

in winter

Flying

I.D.: L: 4.1–5.3" WS: 12–14" WT: 0.3–0.6 oz. A medium to large bat with enormously long ears (about ⅔ of the body length). There is a unique flap of skin projecting from the base of each ear over the nose. There is also a tuft of light-colored hairs at the base of each ear. The hairs on the back are black at the base and vary from light tan to almost black at the tip. The belly is slightly lighter than the back.

In Flight: Flies about 30 feet above the ground and makes diving swoops in meadows. Flight is swifter than that of most bats. It is a very adept flier and is even able to hover and fly vertically. In open areas, the bat has a fast and direct flight. Can emit loud peeps when flying.

Feeding: Emerges from roost only in complete darkness. Feeds in forested areas of the Southwest, where it gleans insects from the vegetation. Commonly found in pine-oak forested canyons around pinyon pines. Can fly between, within, and below tree canopies. May also be found in nonforested, hot, and humid areas like Mojave Desert scrub and in open areas.

Foods: Eats mostly small moths, but may also eat a variety of beetles, roaches, and flying ants.

Echolocation Frequency: 16–25 kHz.

Roosting: Roosts in colonies in rocks, caves, mines, and under loose bark of dead trees. When roosting, its long ears lie flat along its back. The ears perk up when the bat is alert and ready to take flight.

Migration/Hibernation:
Migrational movements of Allen's Big-eared Bats are largely unknown. They probably hibernate for short intervals in high-elevation areas, and at low to middle elevations are active throughout the winter. They probably use caves and mines as hibernating sites.

Breeding. Forms maternity colonies of pregnant females in rock shelters, mines, and trees. One pup is born usually during the months of June and July.

Tip Uncommon throughout most of its range. This bat's biology is poorly known.

Silver-haired Bat

Lasionycteris noctivagans

Roosting

in summer

in winter

Flying

I.D.: L: 3.6–4.6" WS: 11–13" WT: 0.3–0.4 oz. A medium-sized black bat with silver-tipped fur. The tail membrane is lightly furred closest to body. Wings are black. Dorsal fur has a characteristic silver-tipped color. Can live 12 years or more.

In Flight: One of the slowest-flying bats in North America and could be misidentified as a large moth. Its slow, leisurely flight makes it recognizable. Can be seen flying singly or in pairs, occasionally in groups of 3–4. Silver-haired Bats are occasionally seen flying before the sun has completely set. Fishermen have reported catching this bat on their hooks. Presumably, the bat mistakes the lure for food.

Feeding: Emerges around sunset or shortly before. Usually feeds over ponds and streams, but also forages above treetop level in woods. It sometimes can be seen flying the same pattern each night.

Foods: Feeds on insects — mostly moths, but also true bugs, flies, mosquitoes, termites, and beetles.

Echolocation Frequency: 26–38 kHz.

Roosting: Found roosting singly or in small groups in wooded areas. Typically roosts in hollows and cracks and crevices of trees. Less commonly found roosting in old woodpecker holes and beneath rocks. Roosts are usually between 3 and 16 feet above the ground.

Migration/Hibernation: Silver-haired Bats are migratory and apparently travel in groups. During migration, the bats can be found in open sheds, garages, and outbuildings. They can also be seen resting in lumber piles, railroad ties, and fenceposts. Once on hibernation grounds, Silver-haired Bats hibernate in trees, buildings, rock crevices, caves, or any other area that is protected from the winter weather. In northern areas some bats may overwinter; however, appropriate shelters must be available.

Breeding: Mating occurs in fall, and the sperm is stored in the female's body until spring. These bats may occasionally form nursery colonies, with males and females roosting separately. Gestation is 50–60 days. Offspring are generally twins, usually born in June or July. Pups are born without fur, have pinkish skin and mottled tan-and-black wings, and are born with eyes closed. Newborns weigh roughly 0.1 ounce each, and therefore twins make up about 40 percent of the mother's weight. Pups can forage successfully at 4–5 weeks of age.

Tip Fairly common to uncommon in its northern range. Especially found in old-growth forests. Uncommon in Gulf states, Tex., and southern Calif.

Western Red Bat*

Lasiurus blossevillii

Roosting

in summer

in winter

Flying

I.D.: L: 4.06" WS: 11–13" WT: 0.4–0.5 oz. A medium-sized bat with long pointed wings. Its ears are short and round. The tail membrane is fully furred. The color is orange-brown to yellow-brown. One of the most beautiful bats in North America.

In Flight: Long pointed wings and a long tail membrane make it recognizable in flight. They first fly high in the air, where they flutter and fly slowly and erratically. They then may descend to treetop level or as low as a few feet above the ground to feed. At that point they fly swiftly (they have been clocked at 40 miles per hour). They often fly the same course each evening.

Feeding: Emerges early in the evening and feeds around forests, rivers, fields, and urban areas. May also feed around streetlights, like the Eastern Red Bat (p. 66).

Foods: Eats moths and other insects.

Echolocation Frequency: 45–54 kHz.

Roosting: Roosts alone in leaves of large shrubs and trees. Roost trees are in habitats that border forests, rivers, fields, and may even include urban areas. When at rest, these bats sometimes hang by one foot, with the other foot and their head tucked under their furry tail membrane, like the Eastern Red Bat. When roosting this way, they look like a small, red, furry ball.

*Little research has been done on the Western Red Bat because until fairly recently it was not recognized as a species separate from the Eastern Red Bat. Therefore we used the research available on the Western Red Bat but also relied on information on the Eastern Red Bat in writing this section.

Migration/Hibernation: Migratory during certain months, some traveling only short distances. For example, Western Red Bats were found in San Francisco in winter but not in summer. About 60 miles northeast of San Francisco, they were found in spring but not during winter. Others may travel longer distances to hibernating areas.

Breeding: May give birth to up to three pups at a time, in mid-May to late June. Pups are born furred. The mother bat nurses her babies until they are old enough to hunt for insects on their own. Because of the weight of three pups, if the mother has to move the young ones, she may not be able to fly. Occasionally, red bat mothers are found on the ground, clutching their pups but unable to take flight because of the excess weight.

Tip Common throughout its range in the western United States.

Eastern Red Bat

Lasiurus borealis

Roosting

in summer

in winter

Flying

I.D.: L: 3.66–4.61" WS: 11–13" WT: 0.2–0.6 oz. A medium-sized bat with long pointed wings and short rounded ears. This is a beautifully colored bat with bright orange-brown to reddish-brown fur. In the East, males (large photo) are much more brightly colored than females (small photo). Tail membrane is completely furred. One of the most beautifully colored bats in the United States.

In Flight: Often flies before dark, when reddish color can still be seen. The long pointed wings also make these bats recognizable in flight. When flying high in the air, they seem to flutter and fly a slow, somewhat erratic flight. When flying lower (treetop height to ground level), they fly straight or in wide semicircles, and the flight is swift. Red bats have been clocked at speeds of 40 miles per hour. They often fly the same territory each night, and many times fly only 600–1,000 yards from their day roost.

Feeding: Emerges early in the evening. Commonly seen feeding below streetlights, even in suburbs, to take advantage of insects attracted to the light. Also feeds among trees in clearings and over water.

Foods: Eats moths, crickets, flies, true bugs, beetles, and other types of insects.

Echolocation Frequency: 39–50 kHz.

Roosting: A solitary bat that roosts hidden in leaves, especially on the south side of the tree. They usually hang from the petiole (stem) of a leaf, but will also hang from a twig or branch. Red bats commonly hang by one foot when roosting and look much like a dead leaf. The tail membrane wraps around the body, almost like a blanket. Preferred roost trees provide shade and cover above and to the side but are free of leaves below. Roosts are usually 4–10 feet above the ground and generally near streams, fields, and urban areas.

Migration/Hibernation: In colder areas of their range, Eastern Red Bats may migrate in groups to more southern states and hibernate in trees. They can withstand subfreezing temperatures when hibernating. Occasionally may be seen flying on warm winter days.

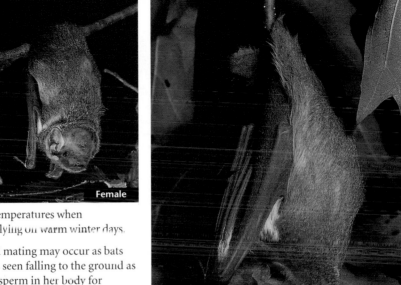

Female

Male

Breeding: Mates in Aug. and Sept., and mating may occur as bats are flying. There are reports of red bats seen falling to the ground as they are mating. The female stores the sperm in her body for several months. Fertilization occurs in spring, and gestation is 80–90 days. One to five pups are born (the largest litter sizes ever reported for any species of bat), furless and with eyes closed; however, pups grow peach fuzz within several days. Occasionally, mother red bats have been found grounded, with young clinging to them. In this circumstance, the mother may have been trying to move her young and the extra weight made her too heavy to fly.

Tip Common throughout its range. Especially common in northern states during summer months.

Hoary Bat
Lasiurus cinereus

Roosting

in summer

in winter

Flying

I.D.: L: 5.1–5.9" WS: 13–16" WT: 0.7–1.2 oz. A large and beautifully colored bat. Blackish-brown or tan color at base of fur; tips are white, giving the bat a frosted appearance. Tail membrane fully furred. Ears short and rounded, with black edges. This is one of the most widespread species in the U.S., with an endangered subspecies found even in Hawaii. Not usually confused with other bats.

In Flight: The flight is swift and direct but not maneuverable. This swift flight and its large size make the Hoary Bat recognizable in flight. Regularly makes a chattering sound, which can be heard by human ears, during flight. Flies mostly in uncluttered areas. In southern states the Hoary Bat will fly in colder temperatures and warm winter afternoons.

Feeding: Usually emerges late in the evening. Hunts at treetop levels in open areas like clearings, fields, and over streams. May also be attracted to insects at outdoor lights. When insects are scarce, the Hoary Bat may defend its feeding territory. May forage in groups.

Foods: Preference for moths. Will also eat true bugs, mosquitoes, dragonflies, and other insects.

Echolocation Frequency: 21–32 kHz.

Roosting: A solitary bat that roosts in foliage of trees. Roosts are usually 7–20 feet above the ground and are leafed above but open below. Roost trees are usually at the edge of a clearing. The Hoary Bat blends in well with the bark of a tree.

Migration/Hibernation: Bats in northern states make long migrations to warmer states, where they may also use trees. Like many types of birds, Hoary Bats have been observed migrating together in flocks. When resting during long migrations, the bats may roost together in groups of 2–7, especially if roosting sites are scarce.

Breeding: Mating occurs in late summer, females store sperm in their bodies for several months, and fertilization occurs in spring. Gestation is 60–90 days. Litter size is most often two. Pups are born in May, June, or July. Babies are born with eyes closed and have

short, fuzzy, silvery fur on back and a naked belly. Newborn Hoary Bats weigh about 0.2 ounce. The mother leaves pups in roost as she forages for insects. Eyes open at 12 days, and by 33 days the young can already fly. Even after young can fly, mother and pups stay together for several weeks longer.

Tip Has a widespread distribution, but not typically common. Can be seen around streetlights and in coniferous areas in summer, especially in the Southwest.

Southern Yellow Bat

Lasiurus ega

Roosting

in summer

in winter

Flying

I.D.: L: 4.02–4.65" WS: 13–15" WT: 0.4–0.5 oz. A large bat with yellow to tan fur. It has long wings and short rounded ears. Tail membrane is long and only furred nearest its body. Belly fur extends onto the underside of the wing membrane.

In Flight: Flies slowly and steadily across clearings, but may also fly at high speeds. These bats have been recorded flying 75 feet above the ground in a straight line.

Feeding: Emerges at dusk. Feeds on insects over clearings and over streams, ponds, rivers, and swimming pools. Feeds mostly from sundown to 5 hours after sunset.

Foods: Consumes small to medium-sized insects at night. It appears to be an efficient feeder; when caught in a mist net only 2 hours after dusk, the bat already has a full belly.

Echolocation Frequency: Unknown.

Roosting: Tree-roosting bat that lives alone. Roosts are commonly found roughly 15 feet above the ground in well-concealed leafy trees. Most records of these bats in the U.S. have been in palm trees along the Rio Grande near Brownsville, Tex. In parts of its range other than the U.S., these bats are found at lower elevations, below 1,650 feet, in moist habitats and tropical evergreen forests.

Migration/Hibernation: The Southern Yellow Bat may be migratory. One bat was captured on a ship 208 miles off the coast of Argentina. It seems to be a year round resident in Brownsville, because it was found there in December. The Southern Yellow Bat does not hibernate, but it may go into a daily torpor.

Breeding: Pregnant females have been found in April and June. Females may form small maternity colonies when pregnant. They have two to four pups in late April through July. First-year young are capable of breeding.

Tip Rare in the United States, and not much research has been done on this bat.

Northern Yellow Bat

Lasiurus intermedius

Roosting

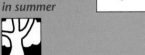

in summer

in winter

Flying

I.D.: L: 4.8–5.2" WS: 14–16" WT: 0.5–0.7 oz. A large bat with a yellow to tan color. Has an overall yellowish hue about it. Tips of fur may be gray or brown. Wings are long, and tail membrane is only furred closest to body. Short rounded ears give it an elflike appearance.

In Flight: Look for several bats emerging in the evening from a tree with Spanish moss. Usually flies 17–23 feet above the ground, at tree canopy level or above.

Feeding: Forages for food in very open grassy areas. Examples of favorite foraging sites include airports, open pastures, golf courses, and bodies of water. Northern Yellow Bats may be seen feeding in groups during the summer months. Males are not usually in these feeding aggregations.

Foods: Beetles, true bugs, flies, mosquitoes, and other insects are eaten.

Echolocation Frequency: 29–41 kHz.

Roosting: Northern Yellow Bats live and raise young in close association with Spanish moss. They are mostly solitary but also are somewhat colonial, found in small numbers and not choosing to cluster together. A single tree with Spanish moss may house several yellow bats. Roost trees are usually in wooded areas with water nearby. These bats have also been reported living among leaves of tall (over 20 feet) palm trees, where their droppings were seen on the ground and loud bickering noises were heard during the day. Yellow bats are not abundant unless their roost trees are near grassy areas in which they forage.

Migration/Hibernation: Probably do not migrate and may roost together in palm trees in winter. They may emerge on warm evenings during winter months.

Breeding: Mating occurs in fall and winter. Sperm is stored in the female's body for several months. Pregnant females may form maternity colonies. Two to four pups are born in May, June, or July. Babies weigh roughly 0.11 ounce. Mother bats leave pups as they forage for insects at night. Young are reared in Spanish moss.

Tip Fairly common throughout its range. May be the most abundant bat in some parts of Florida.

73

Seminole Bat
Lasiurus seminolus

Roosting

in summer

in winter

Flying

I.D.: L: 3.5–4.5" WS: 11–13" WT: 0.3–0.5 oz. A medium-sized bat with deep mahogany–colored fur. Sometimes fur is tipped with silver. Could be confused with the Eastern Red Bat (p. 66), but the Seminole Bat's fur is a much darker, richer color. Its wings are long and pointed, and the ears are small and rounded. The tail membrane is completely furred. Males and females look alike.

In Flight: Flies high, above treetop level, 20–45 feet above the ground, during dark nights. When the moon is bright Seminole Bats will fly below the tree canopy. The flight pattern is swift and seems to be direct. These bats can be seen flying during all months of the year, but the temperature needs to be 70 degrees Fahrenheit or above for them to become active.

Feeding: Emerges early in the evening. Insects are captured in and around tree canopies, over water, and in clearings. Seminole Bats have also been seen capturing insects at streetlights.

Foods: Eats many types of insects, including true bugs, mosquitoes, beetles, flies, and crickets. The bats have been observed flying repeatedly into a tree canopy where swarms of insects are flying.

Echolocation Frequency: Unknown.

Roosting: Roosts singly. During fall, winter, and spring, usually roosts in the middle of Spanish moss on the southwestern side of a tree; during summer, usually roosts in pine trees. The roost is usually high enough that the bat can drop down into a clear area when taking flight. Therefore the roost trees are usually bordering a clearing. Seminole Bats have also been reported roosting in caves and beneath the bark of trees. Although they are solitary bats, several bats sometimes occupy the same roosts on different days. When the bat is resting, its furry tail membrane is folded down over its belly.

Migration/Hibernation: In the northern part of their range, Seminole Bats migrate south to warmer states. In southern states, they probably do not migrate. These bats probably do not hibernate. Bats have been seen during winter months in southern states flying on warm nights. Seminole Bats probably roost in trees during the winter months.

Breeding: Pups are born in late spring to early summer. One to four babies are born, the average being three. Researchers have noted occurrences of Seminole Bats in late summer outside their normal breeding range. This indicates the bats wander after their young are weaned.

Tip In some parts of the South it may be the most abundant bat seen flying.

75

Western Yellow Bat

Lasiurus xanthinus

Roosting

in summer

in winter

Flying

I.D.: L: 4.0–4.7" WS: 13–15" WT: 0.4–0.5 oz. A large bat that closely resembles the Southern Yellow Bat (p. 70). Until fairly recently, it was actually listed as the same species; however, studies have shown it is genetically different. The fur is a yellowish hue with gray tips. Wings are long and pointed, and the ears are short. Tail membrane is furred close to the body only.

In Flight: No information available; see Southern Yellow Bat for flight information that may compare to Western Yellow Bat.

Feeding: No information available; see Southern Yellow Bat for feeding information that may be relevant.

Foods: Eats many types of small to medium-sized insects that fly at night.

Echolocation Frequency: 33–47 kHz.

Roosting: Has been found roosting in hackberry and sycamore trees in N.Mex. and in palm trees in Calif., Ariz., and Nev.

Migration/Hibernation: Several have been found hibernating in dead fronds of palm trees in Ariz. during winter months. This indicates that the bats do not migrate; however, they do hibernate.

Breeding: Number of pups ranges from two to four, with two the most common. Young are born in June.

Tip Common only in south-central Arizona; uncommon everywhere else in its range.

Southwestern Bat

Myotis auriculus

Roosting

in summer

in winter

?

Flying

I.D.: L: 3.4–4.0" WS: 10–12" WT: 0.2–0.3 oz. Small bat with a brownish color. Fur is dull, not glossy. Ears are long, and ears and wings are brown.

In Flight: Flies mostly around areas of rocky cliffs where water is available. Has been clocked flying up to 12 miles per hour.

Feeding: May land on buildings and tree trunks to capture insects, but mostly forages in the sky. Most active feeding about 1½–2 hours after sunset.

Foods: Usually eats moths with 1–2 inch wingspans. Males tend to eat more moths than females do.

Echolocation Frequency: 35–60 kHz.

Roosting: Probably colonial. Roosts have been found in rocks and crevices. Night roosts have been found in buildings, mines, and caves. When found in night roosts, colony size was less than three bats at a time.

Migration/Hibernation: Bats are thought to migrate.

Breeding: One pup, usually born in June or July.

Tip Common in pine forests; however, not much Information is available on this species of bat.

Southeastern Bat
Myotis austroriparius

Roosting

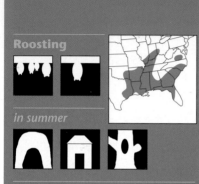

Roosting

in summer

in winter

Flying

I.D.: L: 3.03–3.51" WS: 9–11" WT: 0.2–0.3 oz. A small bat with brownish to grayish fur on its back; in parts of its range is reddish in color. Tan to whitish fur on its belly. Hair on toes is long, extending beyond the nails.

In Flight: Slow acrobatic flight. Usually seen flying over water.

Feeding: Emerges late in the evening and flies directly to a nearby water source, such as a pond or stream, to feed. When feeding, it flies very low, just over the water's surface, to catch insects.

Foods: Specializes in small, soft-bodied aquatic insects, such as midges and caddis flies.

Echolocation Frequency: 40–80 kHz.

Roosting: Females use caves during the maternity season. These caves usually have water sources within them or are close to water. In the South, this bat may also use buildings, mines, and hollow trees, along with caves, as roost sites. Males usually roost alone or in small bachelor colonies separated from females.

Migration/Hibernation: Southeastern Bats are not known to migrate. In winter they often leave caves to roost in small groups in outdoor areas. They are often found roosting over water, in places that are protected from the severe cold weather but are cooler than caves. Some winter roosts include bridges, storm sewers, boathouses, buildings, and hollow trees. In northern states the bats will hibernate in these roosts; in southern states they stay fairly active all winter long.

Breeding: Females have two pups between April and May. The Southeastern Bat is the only *Myotis* bat that regularly gives birth to more than one pup. Young bats roost in clusters separated from adults during the day. In the evening, the mother leaves her young to begin feeding. Young can fly at 5–6 weeks of age. Nursery colonies disperse in Oct. for their winter roosts.

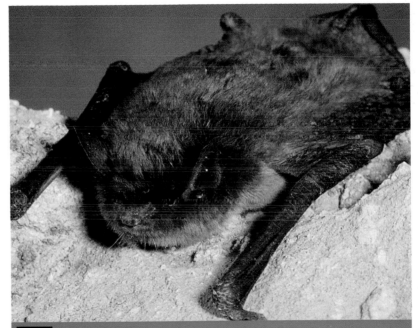

Tip Once common, but because of human persecution and habitat destruction, populations are decreasing quickly.

81

California Bat
Myotis californicus

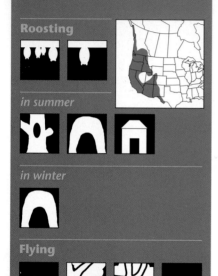

Roosting

in summer

in winter

Flying

I.D.: L: 2.8–3.7" WS: 11–13" WT: 0.1–0.2 oz. A small bat with light tan to black fur. Feet are tiny. Fur is dull, not glossy. Ears are medium-sized. One of the smallest of all our U.S. bats. Lives 15 years or more.

In Flight: Flight is slow and erratic and usually low (5–10 feet) above the ground. Cannot fly well when windy. Flies in set foraging paths (usually loops) about 8–10 feet above the ground.

Feeding: Emerges shortly after sunset to begin feeding on insects. Flies mostly over water (near the surface) and among forest edges. Can sometimes be seen foraging in tree canopies and around tree clumps (high up in some areas) and in open areas. It locates a concentration of insects and slowly maneuvers in and around them to feed. Flies slowly with frequent abrupt changes in direction when pursuing insects. Hunts and feeds rapidly, then rests. Forages for shorter periods when weather is cold. Does not seem to be territorial about feeding grounds.

Foods: Feeds on small insects such as flies, moths, and beetles.

Echolocation Frequency: 48–81 kHz.

Roosting: Females and males roost separately from each other during summer. Females roost alone, or form small maternity colonies, under bark or in hollow trees, caves, mines, and buildings.

Migration/Hibernation: California Bats have been found hibernating in small numbers in caves and mines in high-elevation areas. In other areas, they may stay fairly active all winter long, but for much shorter periods of time each night. Females and males are found roosting together during winter months.

Breeding: Mating occurs in fall, and the sperm is stored in the female's body until spring. One pup is usually born in July, and it can fly by the time it is about 1 month old

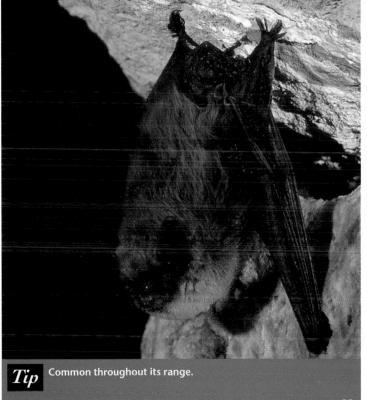

Tip Common throughout its range.

Western Small-footed Bat

Myotis ciliolabrum

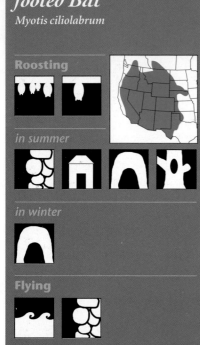

Roosting

in summer

in winter

Flying

I.D.: L: 3.0–3.6" WS: 8–10" WT: 0.1–0.2 oz. Very small bat. Fur is usually brown but can vary from light brown to yellowish color on back and is paler on belly. Ears are long. Ears, wings, and tail membranes are black. The face is also black and contrasts with pale fur, making the bat look as though it is wearing a mask. Has been documented living up to 12 years.

In Flight: Flies slowly and somewhat erratically. Usually flies 3–10 feet above the ground. Is usually seen flying in areas at middle to high elevations.

Feeding: Emerges at dusk. May forage over water and along cliffs and rocky slopes.

Foods: Eats a variety of insects, including caddis flies, flies, beetles, and moths.

Echolocation Frequency: 41–77 kHz.

Roosting: Roosts alone or in small groups, in rock faces and clay banks where it can find crevices to crawl into. Has also been found roosting in barns, between boulders, in caves and mines, and beneath bark. It prefers small, tight, warm crevices.

Migration/Hibernation: Western Small-footed Bats hibernate in caves and mines. They usually hibernate alone, wedging themselves into tight crevices with their bellies against the ceiling and their heads downward.

Breeding: Mating probably occurs in fall, before bats begin hibernation. Sperm is stored in the female's body until spring. Young are born from June to July. Females usually have only one pup. Newborn bats are extremely small. The young are able to fly at 1 month of age.

Tip Uncommon in most areas; however, more information is needed to determine status of this bat.

Western Long-eared Bat

Myotis evotis

in summer

in winter

I.D.: L: 3.43–3.94" WS: 10–12" WT: 0.2–0.3 oz. Small bat with brown fur on back and lighter fur on belly. Fur is long and glossy. Ears and wings are black and usually contrast with paler fur. Ears are very long. Has been reported to live up to 22 years.

In Flight: This bat flies slowly but maneuvers well and is able to hover. When flying over water, it tends to fly fast. Flies nearer to the ground in cooler temperatures.

Feeding: Emerges 10–40 minutes after dusk to forage around treetops and over water in forested areas. Feeding behavior is flexible. It can eat insects in the air or pick them off vegetation or the ground. This feeding behavior may enable the bat to successfully survive in high cool altitudes, where flying insects are scarce.

Foods: Seems to prefer moths and beetles. Other prey items include true bugs, flies, and other insects, as well as spiders. Males eat more moths than females.

Echolocation Frequency: 37–77 kHz.

Roosting: Small maternity colonies of 12–30 individuals are formed in summer months. Males and nonpregnant females roost singly or in small groups, sometimes roosting near the maternity colony but not in it.

Roosting sites include abandoned buildings, hollow trees, niches under bark, caves, mines, cliff crevices, and even sinkholes.

Migration/Hibernation: Probably migrates short distances. A few individuals have been found hibernating in caves and mines. One member of this species was found hibernating in a garage. However, winter records are poorly documented.

Breeding: Mating probably occurs in fall or early winter, and fertilization is delayed until spring. Females have one pup, which is born in June or July.

Tip Thought to be moderately common to uncommon throughout its range. More information is needed to determine status of this bat.

Gray Bat

Myotis grisescens

Roosting

in summer

in winter

Flying

I.D.: L: 3.2–3.8" WS: 11–13" WT: 0.2–0.6 oz. A small bat (it is, however, large for a *Myotis*) with gray-colored fur; hairs are uniformly gray from base to tip. Wing membrane is attached to the ankle of the foot instead of to the base of the toes, as in other *Myotis* species. Lifespan is 14–16 years or more.

In Flight: Flight speeds average 12.6 miles per hour. Gray Bats may travel up to 32 miles to forage for insects.

Feeding: Forages mostly over rivers and lakes. The Gray Bat also feeds in the protection of forest canopies, many times below treetop height.

Foods: Moths, beetles, flies, and mayflies are the most common prey, but Gray Bats may eat other insects as well. A colony of about 250,000 Gray Bats can eat a ton of insects each night.

Echolocation Frequency: Unknown.

Roosting: Roosts almost exclusively in caves. Females form maternity colonies of hundreds to thousands of bats. In summer found mostly in caves with streams and high ceilings that trap warm air. Because of these special requirements, fewer than 5 percent of all caves are suitable for the bats. Gray Bats are very prone to disturbance and take flight easily, even when just a flashlight is turned on the colony.

Repeated disturbances cause them to move or to leave the cave altogether. Cave explorers need to be very careful not to disturb these vulnerable bats.

Migration/Hibernation: These bats use different caves in summer and in winter. In hibernation, the Gray Bat's forearms may stick out at an angle instead of being held close to the body as with some other species. In fall the bats leave their summer roosts and migrate to winter caves, which tend to be vertical and have cool rooms suitable for hibernation.

Breeding: Mating occurs in Sept. and Oct., and the sperm is stored in the female's body for several months. In spring, females become pregnant. One young is born in May, June, or July. Mothers and pups form great masses on cave ceilings. Young can fly 20–25 days after birth. Pups are born naked, with eyes closed, weighing 0.08 to 0.11 ounce.

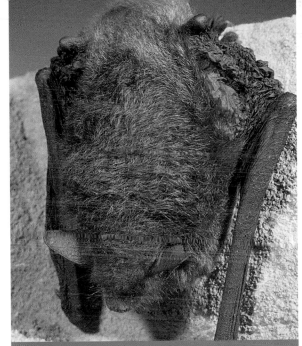

Tip Endangered. Most of these bats roost in a small number of caves in the U.S., making them vulnerable to habitat destruction. Populations have declined 50 percent since 1965.

Keen's Bat
Myotis keenii

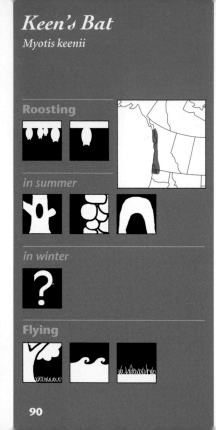

Roosting

in summer

in winter

?

Flying

I.D.: L: 1.6–2.2" WS: 8–10" WT: 0.1–0.2 oz. A small bat with brown glossy fur. Ears are long (almost 1 inch). Ears and wings are dark brown.

In Flight: Flies slowly and usually high. Maneuvers quite well.

Feeding: Emerges just after dark. Hunts high, along edges of forests, over ponds, and in clearings. Catches insects in the air.

Foods: Eats small insects — moths, for example — on the wing.

Echolocation Frequency: Unknown.

Roosting: Roosts singly or in small maternity colonies in tree cavities, rock crevices, and small caves. Usually found in coastal forest habitats.

Migration/Hibernation: Unknown.

Breeding: Mating probably occurs in autumn. One pup is born in June or July.

Tip Status is unknown, although probably rare.

Eastern Small-footed Bat

Myotis leibii

Roosting

in summer

in winter

Flying

I.D.: L: 2.9–3.2" WS: 8–10" WT: 0.1–0.2 oz. An extremely small bat, one of the two smallest in the U.S., weighs about as much as a quarter. Fur is golden brown and glossy on its back and lighter on belly. Ears and wings are black. Has a black mask on face. Feet are small. Can live 12 years or more.

In Flight: Flies slowly and erratically 3–10 feet above the ground.

Feeding: Emerges shortly after sunset. Although little is known about the feeding behavior of this bat, it has been seen foraging over water.

Foods: Eats beetles, true bugs, flies, leafhoppers, flying ants, and other insects.

Echolocation Frequency: Unknown.

Roosting: Females form small colonies of fewer than 20 bats. Roosts are commonly in crevices of cliffs, but can also be beneath tree bark, in buildings and bridges, and in rock piles. Male bats may roost alone.

Migration/Hibernation: Hibernates in mines and caves only a short distance (less than 25 miles) from summer roost. It chooses narrow cracks and crevices when roosting. When hibernating, it roosts by itself, although up to 400 Eastern Small-footed Bats may occupy the same cave. It is a very hardy bat, especially for its small size, and can withstand cold temperatures. It is most commonly seen near the entrance of the mine or cave, where temperatures are coolest. It is one of the last bat species in North America to begin hibernation.

Breeding: Mating occurs in late summer or early fall, and the sperm is stored in the female's body for several months. In spring, females become pregnant. One pup is born in June.

Tip Uncommon; some authorities consider it one of the rarest bats in the United States.

Little Brown Bat
Myotis lucifugus

Roosting

in summer

in winter

Flying

I.D.: L: 2.4–4.0" WS: 9–11" WT: 0.2–0.5 oz. A small bat with glossy pale tan to dark brown fur. Coat is evenly colored. Long hairs on toes extend past toenails. Ears are small and black. This bat can live 34 years or more, with males living longer than females.

In Flight: Usually repeats its flight pattern when flying. When these bats approach their roost at dawn, they may make many passes around roost before entering.

Feeding: Emerges at late dusk. Little Brown Bats often forage over water, flying low over the surface. They may also forage around trees and over lawns, pastures, or streets near the roost. They can forage at heights of 10–20 feet over open areas. Insects are usually caught in the bat's wing, transferred to the tail membrane, then grabbed with the teeth.

Foods: Eats gnats, crane flies, beetles, moths, and mayflies. Little Brown Bats can sometimes eat more than their entire body weight in insects each night.

Echolocation Frequency: 38–62 kHz.

Roosting: Hundreds or thousands of these bats may roost in hot attics or other buildings. They are likely to use bat houses and manmade structures and also use dead and dying trees. Roosts are usually near bodies of water. Female Little Brown Bats often return to the same roost each year.

Migration/Hibernation: Migrates to winter roosts in fall. Hibernates in caves and mines in groups of up to 300,000 bats. Leaves for summer roosts in spring.

Breeding: Mating occurs in fall, and the sperm is stored in the female's body for several months. In spring, females become pregnant. Gestation is 60 days. One pup is born in May, June, or July. They are born furless and with eyes closed. Young bats roost under their mother's wing during the day and are left at the roost as she forages at night. Little Brown Bats are able to fly at 14 days old and are adult size at 21 days.

Tip Very common bat in northern United States and in Canada; scarce in southern United States.

Northern Long-eared Bat
Myotis septentrionalis

in summer

in winter

Flying

I.D.: L: 3.2–3.8" WS: 9–11" WT: 0.2–0.4 oz. A small bat with pale to dark brown fur. Ears are very large (about 0.6–0.7 inch) and almost black. Wings and tail membrane are almost black as well. This bat can live more than 18 years.

In Flight: Flies mostly in forested areas.

Feeding: Emerges soon after dusk and feeds mostly in the first 2 hours after taking flight. Forages for insects under forest canopies, over water, along paths and roads, and near edges of forests. Eats insects in the air, but also picks them off leaves and the ground, which allows it a broader diet. It finds insects on the ground by listening to their movements with its large ears. Insects are then carried to a night roost to be consumed.

Foods: Moths, beetles, caddis flies, true bugs, and other insects are consumed.

Echolocation Frequency: 43–77 kHz.

Roosting: Males roost alone during summer months. Females roost together in small colonies of fewer than 60 individuals and prefer to roost in tight crevices. They have been found roosting in barns, under shutters or shingles, but most often roost in tree hollows or under the bark.

Migration/Hibernation: Bats leave for hibernating sites in Aug. and Sept. Winter roosts are usually fairly close to summer sites. When first arriving, large numbers of bats fly around the entrance of a cave or mine to mate. When hibernating, they roost alone or with a few others. The entire colony size is usually less than 300 bats, however some colonies of more than 1,200 have been reported. These bats are rarely seen hibernating because they are in cracks in the walls. The hibernation period may last 8–9 months in northern areas. A threat to the Northern Long-eared Bat is the sealing of abandoned mines. Closing mines with gates rather than sealing them is a better alternative, to give bats proper hibernating sites.

Breeding: Mating occurs in fall, and the sperm is stored in the female's body for several months. In spring, females become pregnant. Gestation is about 60 days, and one pup is born in June or July. At 30 days old, the young are weaned and able to fly.

Tip Common over much of its range.

Indiana Bat
Myotis sodalis

Roosting

in summer

in winter

Flying

I.D.: L: 2.9–3.9" WS: 9–11" WT: 0.1–0.4 oz. A small bat with dark gray to dark brown, almost black, dull fur. Ears, wings, and tail membrane are pinkish. Hairs on toes do not extend past toenails. This bat can live 14 years or more.

In Flight: Flies direct and at a height of about 40 feet when flying around trees. Has been clocked flying 12.5 miles per hour. Usually does not fly more than 3 miles from roost.

Feeding: Emerges about 25 minutes after sunset. Often flies directly to a water source (a river or stream, for instance) to feed. Also flies around the canopy level and along hedges and open areas. Captures insects in mouth, with wings, or with tail membrane.

Foods: Flies, moths, caddis flies, and mosquitoes are common prey. In some areas female Indiana Bats tend to feed on moths more often when they are lactating, and moths, beetles, and other insects after young are raised.

Echolocation Frequency: 41–66 kHz.

Roosting: Females form maternity colonies and males roost near maternity roost or at hibernation sites. Indiana Bats choose dead or mostly dead trees, where they mainly roost under loose and peeling bark or rarely in cavities. Roost trees can be in wooded streamside habitat or in open areas, usually in full sunlight.

Migration/Hibernation: In Aug., the bats migrate from their summer roosts to caves or mines. The bats swarm outside the caves in large numbers, mating, before entering to begin hibernation. Hibernating bats form tight irregularly shaped clusters. These clusters form at the same spots each year, and their locations can be recognized by the brown staining of feces.

Breeding: Mating occurs in fall, and the sperm is stored in the female's body for several months. In spring, females become pregnant. Gestation is roughly 60 days. One pup is born in June, and by July it is able to fly.

Tip Considered endangered and rarely seen. Fewer than 360,000 are still alive.

Fringed Bat
Myotis thysanodes

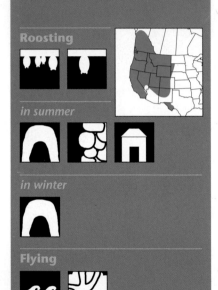

Roosting

in summer

in winter

Flying

I.D.: L: 3.2–4.0" WS: 11–13" WT: 0.2–0.4 oz. A small bat with long ears (0.6–0.8 inch). Color is reddish brown to dark brown on back and paler on belly. Fur is long. This bat has a conspicuous fringe of hair on the outer edge of its tail membrane.

In Flight: Bats fly close to the canopy of trees. Their flight is slow and agile, and they are able to maneuver very well. Will fly on windy and rainy nights. You may see this bat flying around its day roost when young are still unable to fly.

Feeding: Emerges from roost 1–2 hours after sunset and forages along streams and rivers. Most prey are captured in the air, but parts of flightless insects have been found in feces, making it likely this bat also gleans insects from foliage.

Foods: Beetles, moths, flies, leafhoppers, lacewings, and crickets are common food items.

Echolocation Frequency: 26–61 kHz.

Roosting: Roosts in caves, mines, rock crevices, or buildings. May change roosts periodically. This bat roosts with others of the same species. Although colonies of up to 1,200 females and their young have been found, most roost in small clusters. Males usually roost alone and separately from females.

Migration/Hibernation: Known to migrate, but little else is known. It has been found hibernating singly in caves. Researchers have speculated that these bats may move short distances to lower elevations or to warmer areas where they can stay somewhat active throughout the winter.

Breeding: Mating occurs in fall, and the sperm is stored in the female's body for several months. In May, females become pregnant. One pup is born in June or July. Pups are left in clusters separated from adults during the day. Females periodically find their young to nurse them. By 3 weeks of age, the young are adult size and able to fly.

Tip Not usually common.

Cave Bat
Myotis velifer

Roosting

in summer

in winter

Flying

I.D.: L: 1.7–2.2" WS: 11–13" WT: 0.3–0.5 oz. A small bat with fur color ranging from light brown to almost black. Ears are small. Lifespan can be up to 12 years.

In Flight: Has a strong, direct, and less erratic flight than other species of *Myotis*. It has a low maneuverability. Usually flies directly to streams and other water sources at a height of 15–20 feet, or treetop level, to drink before beginning to feed. Can fly 13–15 miles per hour.

Feeding: Emerges before dark and fills its stomach within a half hour. It then rests at a night roost for a few hours before foraging again. This bat forages over desert, floodplains, and water. May fly back and forth over foraging grounds 50–70 yards long. When foraging among sparse vegetation, it flies close to the foliage. May be seen foraging beneath streetlights in desert villages.

Foods: Small moths and beetles are common prey. Diet fluctuates with the season and insect availability.

Echolocation Frequency: 39–66 kHz.

Roosting: Roosts in colonies of 2,000–5,000 in caves and mines, and less often in buildings and under bridges. Frequently roosts with Brazilian Free-tailed Bats (p. 130) and Yuma Bats (p. 106).

Migration/Hibernation: In Kans., Okla., and Tex., this bat is a permanent resident and does not migrate. In Ariz. and Calif., it migrates to winter roosts around Sept. or Oct. and enters caves to begin its hibernation. It will leave again in March for summer roosts.

Breeding: Mating occurs in fall, and the sperm is stored in the female's body for several months. In spring, females become pregnant. Gestation is 60–70 days, and one pup is born in June or July. Young are born with pink furless bodies and brown wings. Mother bats leave young in clusters when foraging at night. At 6–8 weeks of age, the young are able to fly.

Tip Probably fairly common but not often seen.

Long-legged Bat

Myotis volans

Roosting

in summer

in winter

Flying

I.D.: L: 3.0–4.2" WS: 10–12" WT: 0.2–0.4 oz. A small bat with reddish to nearly black fur. Ears are short and round. Underside of wings is lightly furred. Can live 21 years or more.

In Flight: Rapid, strong, and direct flier, capable of flying quickly (has been clocked at 9–11 miles per hour). Flies long distances. Commonly flies 10–15 feet above the ground early in the evening. Later in the night, may fly closer to the ground.

Feeding: Emerges early in the evening and remains active most of the night, even during cool nights. Hunts for insects over water, in forest clearings, around trees and canopies, and near cliffs. May use the same foraging route each night.

Foods: Eats mostly moths, but also consumes flies, termites, lacewings, wasps, true bugs, leafhoppers, small beetles, and spiders. Probably takes advantage of whatever prey sources are available and abundant.

Echolocation Frequency: 42–46 kHz.

Roosting: Abandoned buildings, attics, cracks and crevices, and loose and peeling tree bark are common roosts. Forms large maternity colonies numbering in the hundreds.

Migration/Hibernation: Bats leave summer roosts in fall to migrate to winter roosts in caves and mines. When finding hibernating sites, they fly around in large numbers and mate. There are usually more males than females at these swarming areas.

Hibernation begins in late Sept., and bats form small clusters of individuals as they are hibernating.

Breeding: Mating occurs in late Aug., and the sperm is stored in the female's body for several months. In spring, females become pregnant.

Tip Common in southwest United States.

105

Yuma Bat

Myotis yumanensis

Roosting

in summer

in winter

Flying

I.D.: L: 3.0–3.5" WS: 9–10" WT: 0.1–0.2 oz. A small bat with a light brown to dark brown back and whitish to buffy belly. It has large feet and small ears. Its wings and ears are dark brown.

In Flight: Can fly long distances to feeding grounds. It flies low, rarely above 30 feet from the ground.

Feeding: Emerges early in the evening and typically feeds over water, just above the surface. It is a very efficient feeder and can fill its stomach within the first half hour of feeding. Flies mostly over water, but may also fly over open desert and around streetlights.

Foods: Eats mayflies, caddis flies, midges, beetles, flies, termites, and moths.

Echolocation Frequency: 49–79 kHz.

Roosting: Forms colonies of up to 5,000 bats in buildings, caves, mines, and under bridges. When found in caves, colonies are much smaller. Most roosts are close to water. When maternity roosts are disturbed, females are likely to quickly abandon the roost. Males are solitary and not found near maternity colonies.

Migration/Hibernation: In fall, the bats leave for their hibernating sites, but their destination is unknown. A few have been found in caves. They return to summer roosts in April.

Breeding: Mating occurs in fall, and the sperm is stored in the female's body for several months. In spring, females become pregnant. Maternity colonies are sometimes shared with Little Brown Bats (p. 94). One pup is born in May, June, or July.

Tip Commonly seen along rivers and reservoirs, especially in the Southwest.

Evening Bat

Nycticeius humeralis

Roosting

in summer

in winter

Flying

I.D.: L: 3.3–3.8" WS: 10–11" WT: 0.3–0.5 oz. A small bat with dark brown fur. Ears, wings, and tail membrane are black. Ears are short, and nose lacks fur. Closely resembles a Big Brown Bat (p. 56) but is much smaller.

In Flight: Usually has a medium to fast erratic flight; however, sometimes flies slowly and deliberately. Usually will not be found flying in cold temperatures.

Feeding: Emerges early in the evening. Usually hunts for insects over clearings and ponds; also can be seen near wooded areas. May follow other bats to good foraging grounds; therefore Evening Bats are sometimes seen flying with other species of bats.

Foods: Eats a variety of small insects, including beetles, moths, flies, winged ants, and leafhoppers. Evening Bats are very important for controlling agricultural pests because they eat the adult stage of the corn rootworm (spotted cucumber beetle).

Echolocation Frequency: 35–75 kHz.

Roosting: Large colonies (up to several hundred bats) will roost together in buildings, and smaller colonies (30 or more) may gather under loose bark of trees and in hollow trees, where they form tight clusters. These bats can be seen roosting in towns and rural areas. Males form harems, which they defend from other male Evening Bats. In southern states these bats may form colonies with Brazilian Free-tailed Bats

(p. 130) and Big Brown Bats. Domestic and feral cats kill many of them.

Migration/Hibernation: Probably migrates and hibernates, although Evening Bats have been found in southern states all year round. They are absent in winter in the northern extent of their range.

Breeding: Gives birth in May or June, and usually has two young, but triplets have been documented. Before giving birth, the female will move away from the rest of the colony. Newborns weigh on average 0.07 ounce; their eyes are closed, and their furless bodies are pink with nearly transparent skin. The wings and feet are black. If separated from their mother, the young are very vocal, calling out for her. By the fifth day, most have fur, and they can fly at 20 days old. Juvenile males are expelled from the colony in Aug., Sept., and Oct. These male bats show up in odd places like eaves and outdoor stairways.

Tip Common in southern states but less common elsewhere.

Western Pipistrelle Bat

Pipistrellus hesperus

Roosting

in summer

in winter

Flying

I.D.: L: 2.4–3.4" WS: 7–9" WT: 0.1–0.2 oz. The smallest of all U.S. bats. It has pale fur that varies in color from yellow to light gray to reddish brown.

In Flight: Strong, fast, and erratic flier. Sometimes mistaken for a large moth.

Feeding: Often begins foraging before sunset and feeds throughout the night, even after sunrise, with occasional rest periods. Flies over desert areas, around towns, and above water. Commonly seen over water in arid areas. Usually forages 50–100 feet above the ground. It has been seen foraging with swallows before sunset. The Western Pipistrelle was once viewed being chased away from its feeding ground by a bird.

Foods: Eats caddis flies, stoneflies, moths, small beetles, leaf and stilt bugs, leafhoppers, flies, mosquitoes, ants, and wasps.

Echolocation Frequency: 47–55 kHz.

Roosting: Usually roosts in rock crevices, but may also be found in mines, burrows, buildings, or under rocks. Roosts are usually near water, including swimming pools. This bat is most often found roosting by itself or in small groups. The largest known maternity colony site was 12 females.

Migration/Hibernation: Year-round resident in most areas, and hibernates singly in caves, mines, and rock crevices. Some populations may remain active throughout the winter.

Breeding: Western Pipistrelles usually have two pups in June or July, each weighing less than half an ounce. Young pups begin to fly at about 1 month of age. Mother bats sometimes roost by themselves when rearing young.

Tip Very common in southwest deserts at low elevations and in arid grasslands. May be one of the most common bats seen at dusk in these areas.

Eastern Pipistrelle Bat

Pipistrellus subflavus

Roosting

in summer

in winter

Flying

I.D.: L: 3.0–3.6" WS: 8–10" WT: 0.2–0.4 oz. One of the smallest of all U.S. bats. Its fur is tricolored: the base of the hair is dark, the middle is lighter, and the tip is dark. Color varies from pale yellow to silvery gray to golden brown to black. It can live 15 years or more.

In Flight: Flight is erratic, slow, and weak. It is sometimes mistaken for a large moth. Flies continuously when feeding.

Feeding: Emerges early in the evening to feed, as early as 1 hour before dusk. Foraging area is small. Often feeds over streams and ponds, along forest edges, in pastures, over treetops, and (in the deep South) around Spanish moss. It is never found foraging in deep woods or in open fields unless large trees are nearby.

Foods: Moths, beetles, mosquitoes, midges, true bugs, and ants are common food items.

Echolocation Frequency: 41–50 kHz.

Roosting: Most maternity colonies are found in clusters of leaves in woods. Very few roost in buildings. In the southern states, besides trees, leaves, and buildings, some maternity colonies roost in caves. When pregnant bats are found in buildings, usually 30–40 are clustered together. Smaller numbers, 2–6, are found in trees. Male Eastern Pipistrelle Bats are probably solitary during summer months.

Migration/Hibernation: Migrates short distances (longest known distance was 85 miles) to winter sites. Bats exhibit swarming behavior

outside hibernating areas before entering, thereby mating before the winter months set in. They hibernate in caves, mines, and rock crevices singly and scattered throughout. In the eastern U.S., Eastern Pipistrelles are the species of bat you are most likely to encounter in a cave.

Breeding: Mating occurs in fall, and the sperm is stored in the female's body for several months. In spring, females become pregnant. Young are born in May, June, or July. Usually two pups are born, with closed eyes and a hairless pink body. Young ones still unable to fly make distinctive clicks that their mother uses to find them. Within 1 month, they are able to take flight.

Tip Common throughout most parts of its range.

Rafinesque's Big-eared Bat

Corynorhinus rafinesquii

Roosting

in summer

in winter

Flying

I.D.: L: 3.2–4.3" WS: 10–12" WT: 0.3–0.5 oz. A medium-sized bat with ears over an inch long. Color is gray on back and white on belly. Nose has two lumps on either side. Can live 10 years or more.

In Flight: Remarkably agile, easily flying within and close to vegetation. Flies swiftly, and can almost hover.

Feeding: Emerges late in the evening. Feeds on insects in the air and gleans insects from plants or the ground. Insect prey are sometimes large; therefore this species of bat usually has a night roost, where it brings the large prey to leisurely consume it. Feeds in densely forested areas.

Foods: Eats primarily moths, but will eat other insects, especially when moths are scarce.

Echolocation Frequency: 60–90 kHz.

Roosting: Most often roosts in abandoned buildings in heavily forested areas, hanging in the open. Nursery colonies are rarely in caves. Females form colonies of 6 to several dozen individuals. Males are sometimes found under loose bark and in hollow trees or buildings. When approached, these bats usually rotate their ears, probably trying to keep track of intruders.

Migration/Hibernation: Hibernates in clusters in caves and mines. Awakens frequently during hibernation and moves from one roosting site to another. Often hibernates in well-ventilated caves near the entrance.

Breeding: Mating occurs from fall into winter, and the sperm is stored in the female's body for several months. In

spring, females become pregnant. One baby is born in May or June. Pups are hairless and their eyes are closed when born. By 3 weeks of age, the pups can already fly.

Tip Uncommon.

Townsend's Big-eared Bat

Corynorhinus townsendii

Roosting

in summer

in winter

Flying

I.D.: L: 3.5–4.6" WS: 12–13" WT: 0.3–0.4 oz. A medium-sized bat with huge ears (over an inch long). Fur color varies from pale brown to nearly black on back. Hairs on belly are paler. Has two lumps on either side of nose. This bat can live 16 years or more.

In Flight: A very versatile flier. Can go from swift and agile to slow and hovering. When flying fast, wings sound like a noisy flutter. These bats will usually circle inside caves before exiting for the evening.

Feeding: Usually emerges an hour after sunset. Feeds throughout the night. Usually forages between forested and open areas (edge habitat). When feeding over pastures and other open areas, the bat flies low; when flying near trees, it flies much higher.

Foods: Eats mostly moths, but also preys on lacewings, dung beetles, flies, and sawflies.

Echolocation Frequency: 45–90 kHz.

Roosting: Maternity colonies are usually in warm parts of caves. They may also use old mines, buildings, and rock ledges as roosts. They usually roost from an open ceiling and frequently hang by one foot. Females and young roost in tightly packed clusters of a dozen to up to 1,000 bats. They are vulnerable to disturbance when roosting and may abandon the roost permanently if disturbed.

Migration/Hibernation: Hibernates in clusters of a few bats to up to 100 individuals. Hibernates in caves and mines, near entrances where it is well ventilated. If the temperature gets too

cold, they will move to warmer areas. Ears may be folded against the back or standing erect when hibernating. Bats hibernating by themselves have been seen hanging by one foot. No long-distance migrations are known.

Breeding: Mating occurs from Oct. into winter and as late as Feb. These bats have been seen mating while hanging from the ceiling of a cave. The sperm is stored in the female's body for several months. In spring, females become pregnant. One baby is born in late May, June, or July. Pups are born large, with their eyes closed. Young can fly at 2½–3 weeks of age.

Tip Rarely encountered except occasionally in Southwest. Active management program among state and federal agencies to prevent further decline of this species.

Wagner's Mastiff Bat

Eumops glaucinus

Roosting

in summer

in winter

Flying

I.D.: L: 4.9–6.5" WS: 19–21" WT: 0.9–1.9 oz. A large bat with glossy, dark gray, almost black fur on back. Belly is lighter. The tail is free of the tail membrane. Large round ears bend forward over the head.

In Flight: High fliers. These bats have loud piercing echolocation calls in the audible range and have been heard flying over city streets. Able to fly fast, but not very maneuverable. Fairly easy to recognize because of their large wingspan. Unlike other free-tailed bats, which must drop several yards from a roost in order to fly, the Wagner's Mastiff Bat can take off from horizontal surfaces.

Feeding: Emerges after dusk and usually forages over golf courses, city streets, open areas, swamps, coastlines, and mangroves.

Foods: Beetles, flies, mosquitoes, true bugs, and moths are preyed upon.

Echolocation Frequency: 10–35 kHz.

Roosting: Roosts singly or in small colonies under Spanish tile roofs. Some have been found in royal palm shafts, in low shrubs, near tropical flowers, in chimneys, in old woodpecker nests, and in long-leaf pines. Roosts have been found in both urban and forested areas. Males, females, and juveniles roost together.

Migration/Hibernation: These bats are found year-round throughout their range in southern Florida.

Breeding: One pup is born in June–Sept.

Tip Rare; found only in southern Florida.

Greater Mastiff Bat

Eumops perotis

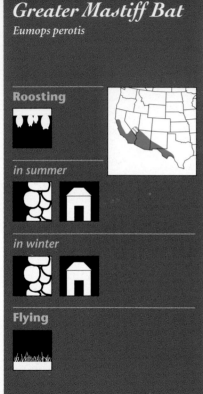

Roosting

in summer

in winter

Flying

I.D.: L: 6.3–7.4" WS: 21–23" WT: 1.6–2.6 oz. The largest bat in the U.S. Fur color on back varies from brownish to dark gray. The belly is lighter. The tail sticks out beyond the tail membrane. Has long narrow wings. Large round ears (1.3–1.9 inches long) bend forward over the head, giving it a bonnetlike appearance.

In Flight: Its large size gives it some difficulty taking flight; therefore it will try to drop about 10 feet from roost to launch into flight. It is able to fly fast and strongly and for long periods of time. Flies high, over 1,000 feet. Sometimes makes a loud high-pitched sound, audible to humans, when flying.

Feeding: These bats chatter loudly before leaving roosts. They emerge only after complete darkness. They forage for insects at very high altitudes. Will fly long distances (15 miles) to search for food. Forage throughout the night and usually in open areas.

Foods: Eats dragonflies, grasshoppers, beetles, true bugs, moths, wasps, and ants.

Echolocation Frequency: 10–18 kHz.

Roosting: Roosts in rock crevices of vertical cliffs and less commonly in buildings. Prefers to roost out of sight in a several-foot-long crevice, and in the furthest part of the crevice, so that its belly and back are touching the rock. Roosts are usually 15 to 20 feet above the ground. These mastiff bats roost in small colonies of fewer than 100 individuals, who move in and about the crevices throughout the day. Males are sometimes found with maternity colonies. The Greater Mastiff Bat will also roost with other species of bats, for example Pallid Bats (p. 54), Big Brown Bats (p. 56), and Brazilian Free-tailed Bats (p. 130).

Migration/Hibernation: These bats do not migrate or hibernate.

Breeding: Males have an oil gland that produces a strong odor to attract females. They mate in early spring. Usually one pup is born in May–Sept. Twins are rare. Pups are furless at birth.

Tip Uncommon throughout its range.

Underwood's Mastiff Bat

Eumops underwoodi

Roosting

in summer

in winter

Flying

I.D.: L: 6.3–6.5" WS: 20–22" WT: 1.4–2.3 oz. A large bat with dark brown fur on back and gray fur on belly. Its tail is free of the tail membrane. Has long narrow wings. The second largest bat in the U.S. Has large round ears that bend forward over the head.

In Flight: It is a strong fast flier. Has been clocked at speeds of at least 26 miles per hour. Able to fly long distances. Can emit a high-pitched sound in flight, so intense it can hurt your ears.

Feeding: Emerges from roost after dusk. Forages in open areas. Has been seen foraging over livestock watering tanks in the desert areas.

Foods: Preys on grasshoppers, leafhoppers, moths, and beetles.

Echolocation Frequency: Unknown.

Roosting: Roosts in high dry places, to allow a free-fall before taking flight. Has been found roosting beneath royal palm leaves and in a large hollow tree. Not much is known about the roosting requirements of the Underwood's Mastiff Bat.

Migration/Hibernation: Nothing known.

Breeding: One pup is born in June or July. Little else is known on reproduction of this bat.

Tip Rare in the United States.

Pallas' Mastiff Bat

Molossus molossus

Roosting

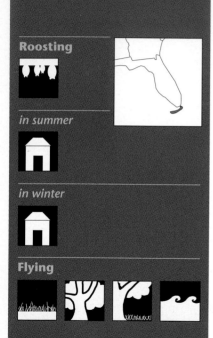

in summer

in winter

Flying

I.D.: L: 4" WS: 11–13" WT: 0.5 oz. A medium-sized bat with a brownish color. Ears are large and rounded. Tail is long and is free from tail membrane. Has a flap of skin below both ears. Snout comes to a point.

In Flight: Flies fast with rapid darting turns similar to the flight of swifts. Very agile fliers. Will fly at heights of above 50 feet and drop into dense foliage for water and foraging opportunities.

Feeding: Usually forages in open areas, above tree canopies, and on the forest edge. Also flies near streams and ponds.

Foods: Eats moths, beetles, and flying ants.

Echolocation Frequency: 30–40 kHz.

Roosting: All colonies in the Florida Keys have been found in the roof crawl spaces of flat-roofed buildings. Females form maternity colonies.

Migration/Hibernation: Apparently does not migrate or hibernate.

Breeding: One pup is born in June–Sept. Babies hang in clusters while mothers forage for food.

Pocketed Free-tailed Bat

Nyctinomops femorosaccus

Roosting

in summer

in winter

Flying

I.D.: L: 3.9–4.7" WS: 13–15" WT: 0.5–0.6 oz. A medium-sized bat with dark brown to gray fur. Tail is long and extends past tail membrane. Wings are long and narrow. A fold of skin near the knees forms a pocket.

In Flight: A swift and nonfluttery flier. It will drop 4–5 feet from its roost before taking flight.

Feeding: Emerges late in the evening, well after dark in some areas. In other areas, like Calif., emerges at dusk. Feeds over stock ponds and other water sources.

Foods: Moths are the most common prey, but beetles, flying ants, flies, leafhoppers, crickets, stinkbugs, lacewings, and grasshoppers are also eaten.

Echolocation Frequency: 18–26 kHz.

Roosting: Roosts in rock crevices in high cliffs. Few have been found in manmade structures. Colonies are usually small, fewer than 100 individuals. You can hear these bats chattering during the day.

Migration/Hibernation: Probably does not migrate. Does not hibernate, but goes into a light torpor, with activity throughout the winter.

Breeding: One pup is born in late June or July. Pups can fly by late August.

Tip Uncommon in the United States.

Big Free-tailed Bat
Nyctinomops macrotis

Roosting

in summer

in winter

Flying

I.D.: L: 5.7–6.3" WS: 17–18" WT: 0.8–1.1 oz. A large bat with glossy light reddish-brown to black fur. Ears are large and tail is free from tail membrane.

In Flight: A fast and strong flier but not maneuverable. It sometimes emits a loud piercing chatter in flight. When approaching roosts in cliffs, it makes steep dives with wings folded and lands many times before entering.

Feeding: Emerges late, when already dark. Has been seen foraging over a large sewer pond and a swimming pool.

Foods: Eats mostly moths, but diet may also include crickets, flying ants, stinkbugs, and leafhoppers.

Echolocation Frequency: 12–21 kHz.

Roosting: Usually roosts in rock crevices in high places. Sometimes found in manmade structures. Roosts in small maternity colonies, with males roosting separately.

Migration/Hibernation: Probably migrates to lower elevations in the fall, but does not hibernate.

Breeding: One pup is born in June or July. Young can fly in Aug.

Tip Found in Utah and at Big Bend National Park in Tex. more than any other places. Otherwise uncommon.

Brazilian Free-tailed Bat

Tadarida brasiliensis

Roosting

in summer

in winter

Flying

I.D.: L: 3.4–4.3" WS: 12–14" WT: 0.4–0.5 oz. A medium-sized bat with dark brown to dark gray fur. The tail is free from the lower half of the tail membrane. Hairs on the toes are very long. Ears are large and rounded and meet the nose, but are not joined. Wings are long and narrow. Can live 15 years or more. Also called Mexican Free-tailed Bat.

In Flight: Can fly at about 40 miles per hour and may fly as far as 25 miles to foraging areas. Flies at various heights, from 15 feet to over 1,000 feet above the ground. Can be recognized by its long narrow wings and its straight and direct flight course. In the Southwest great numbers of these bats can be seen as they are exiting roosts. They look like a long cloud in the sky and sound like the roar of rapids.

Feeding: Emerges before sunset in Southwest and after sunset in Southeast. Usually feeds over meadows, trees, crops, water, and cities.

Foods: Usually eats small moths and beetles. These bats are very important in agricultural areas because they eat many crop pests. A modest estimation of insect consumption is that they eat 6,600 tons of insects each year in Tex. alone. It may be more realistic to triple this amount.

Echolocation Frequency: 24–48 kHz.

Roosting: In the southwestern U.S., these bats roost in caves and can form very large colonies. They can also be found roosting under bridges and in houses, attics, and bat houses. In the Southeast, they do not live in caves; instead they choose manmade structures, living in attics, under roof tiles, and in bat houses. Here they form smaller colonies than in the Southwest.

A characteristic musty odor can be detected near Brazilian Free-tailed roosts.

Migration/Hibernation: In the U.S. Southwest, these bats migrate long distances to Mexico to hibernate in caves. In the Southeast, the bats do not hibernate but move seasonally between roosting sites.

Breeding: Mating occurs in the winter, and females store sperm inside their bodies for several months. In spring, the females become pregnant. Gestation is 77–100 days. A single inch-long pup is born in spring or summer. Its eyes are closed and it has no fur. Young hang with other pups during the day, away from their mothers. Females will nurse them periodically throughout the day and at night when they return from foraging. Babies can fly when they are 5 weeks old.

Tip Commonly seen throughout its range. During summer months, roughly 20 million live in Bracken Cave, Tex., the largest concentration of mammals in the world. Also seen in Carlsbad Caverns, N.Mex.

State Lists

Bats of the U.S. and Canada
By State, Province, or Territory

Listed below are the bats found in each U.S. state and Canadian province or territory. The bats are grouped according to whether they are common or uncommon in each area. "Unknown Status" indicates that no research has been done on the species listed and therefore there is no way to know whether they are common or not.

Bats are listed in the order in which they appear in this book, which matches the generally accepted phylogenetic arrangement.

UNITED STATES

Alabama

Common

- **Big Brown Bat** *(Eptesicus fuscus)*, 56
- **Eastern Red Bat** *(Lasiurus borealis)*, 66
- **Seminole Bat** *(Lasiurus seminolus)*, 74
- **Evening Bat** *(Nycticeius humeralis)*, 108
- **Eastern Pipistrelle Bat** *(Pipistrellus subflavus)*, 112
- **Brazilian Free-tailed Bat** *(Tadarida brasiliensis)*, 130

Uncommon

- **Silver-haired Bat** *(Lasionycteris noctivagans)*, 62
- **Hoary Bat** *(Lasiurus cinereus)*, 68
- **Southeastern Bat** *(Myotis austroriparius)*, 80
- **Gray Bat** *(Myotis grisescens)*, 88
- **Eastern Small-footed Bat** *(Myotis leibii)*, 92
- **Little Brown Bat** *(Myotis lucifugus)*, 94
- **Indiana Bat** *(Myotis sodalis)*, 98
- **Rafinesque's Big-eared Bat** *(Corynorhinus rafinesquii)*, 114

Unknown Status

- **Northern Yellow Bat** *(Lasiurus intermedius)*, 72
- **Northern Long-eared Bat** *(Myotis septentrionalis)*, 96

Alaska

Common

- **Little Brown Bat** *(Myotis lucifugus)*, 94

Uncommon

- **Silver-haired Bat** *(Lasionycteris noctivagans)*, 62
- **California Bat** *(Myotis californicus)*, 82

S T A T E   L I S T S

- **Keen's Bat** (*Myotis keenii*), 90
- **Long-legged Bat** (*Myotis volans*), 104

Arizona

Common

- **Pallid Bat** (*Antrozous pallidus*), 54
- **Big Brown Bat** (*Eptesicus fuscus*), 56
- **Silver-haired Bat** (*Lasionycteris noctivagans*), 62
  Fairly common in forested areas
- **Hoary Bat** (*Lasiurus cinereus*), 68
- **Southwestern Bat** (*Myotis auriculus*), 78
- **California Bat** (*Myotis californicus*), 82
- **Western Small-footed Bat** (*Myotis ciliolabrum*), 84
- **Western Long-eared Bat** (*Myotis evotis*), 86
- **Little Brown Bat** (*Myotis lucifugus*), 94
- **Fringed Bat** (*Myotis thysanodes*), 100
- **Cave Bat** (*Myotis velifer*), 102
- **Long-legged Bat** (*Myotis volans*), 104
- **Yuma Bat** (*Myotis yumanensis*), 106
- **Western Pipistrelle Bat** (*Pipistrellus hesperus*), 110
- **Townsend's Big-eared Bat** (*Corynorhinus townsendii*), 116
- **Greater Mastiff Bat** (*Eumops perotis*), 120
  Fairly common; however heard more than seen

- **Big Free-tailed Bat** (*Nyctinomops macrotis*), 128
  Fairly common; however heard more than seen
- **Brazilian Free-tailed Bat** (*Tadarida brasiliensis*), 130

Uncommon

- **Ghost-faced Bat** (*Mormoops megalophylla*), 42
- **California Leaf-nosed Bat** (*Macrotus californicus*), 44
  May be locally common in some areas
- **Mexican Long-tongued Bat** (*Choeronycteris mexicana*), 46
  One of two species found at hummingbird feeders in S.E. part of state
- **Lesser Long-nosed Bat** (*Leptonycteris curasoae*), 48
  One of two species found at hummingbird feeders in S.E. part of state
- **Spotted Bat** (*Euderma maculatum*), 58
  Fairly common around Grand Canyon in sage-grassland and forest meadow areas
- **Allen's Big-eared Bat** (*Idionycteris phyllotis*), 60
  Uncommon in Mojave Desert areas, fairly common in ponderosa pine forests
- **Western Yellow Bat** (*Lasiurus xanthinus*), 76
  May be common in urban areas planted with ornamental palms
- **Underwood's Mastiff Bat** (*Eumops underwoodi*), 122
- **Pocketed Free-tailed Bat** (*Nyctinomops femorosaccus*), 126

Arkansas

Common

- **Big Brown Bat** *(Eptesicus fuscus)*, 56
- **Eastern Red Bat** *(Lasiurus borealis)*, 66
- **Seminole Bat** *(Lasiurus seminolus)*, 74
  Status unknown, but probably common
- **Southeastern Bat** *(Myotis austroriparius)*, 80
- **Little Brown Bat** *(Myotis lucifugus)*, 94
  Common in northern Ark.
- **Evening Bat** *(Nycticeius humeralis)*, 108
- **Eastern Pipistrelle Bat** *(Pipistrellus subflavus)*, 112
- **Rafinesque's Big-eared Bat** *(Corynorhinus rafinesquii)*, 114
  Fairly common in southern third of state
- **Brazilian Free-tailed Bat** *(Tadarida brasiliensis)*, 130

Uncommon

- **Silver-haired Bat** *(Lasionycteris noctivagans)*, 62
- **Hoary Bat** *(Lasiurus cinereus)*, 68
- **Gray Bat** *(Myotis grisescens)*, 88
- **Eastern Small-footed Bat** *(Myotis leibii)*, 92
- **Northern Long-eared Bat** *(Myotis septentrionalis)*, 96
  Common in the mountainous forested portions of the state;
  uncommon in agricultural areas
- **Indiana Bat** *(Myotis sodalis)*, 98
- **Townsend's Big-eared Bat** *(Corynorhinus townsendii ingens)*, 116
  Ingens indicates this is the subspecies Ozark Big-eared Bat

California

Common

- **Pallid Bat** *(Antrozous pallidus)*, 54
- **Big Brown Bat** *(Eptesicus fuscus)*, 56
- **Silver-haired Bat** *(Lasionycteris noctivagans)*, 62
- **Western Red Bat** *(Lasiurus blossevillii)*, 64
- **California Bat** *(Myotis californicus)*, 82
- **Western Long-eared Bat** *(Myotis evotis)*, 86
  Moderately common to uncommon
- **Little Brown Bat** *(Myotis lucifugus)*, 94
- **Cave Bat** *(Myotis velifer)*, 102
- **Long-legged Bat** *(Myotis volans)*, 104
- **Yuma Bat** *(Myotis yumanensis)*, 106
- **Western Pipistrelle Bat** *(Pipistrellus hesperus)*, 110
- **Brazilian Free-tailed Bat** *(Tadarida brasiliensis)*, 130

Uncommon

- **California Leaf-nosed Bat** *(Macrotus californicus)*, 44
- **Mexican Long-tongued Bat** *(Choeronycteris mexicana)*, 46

- ■ **Spotted Bat** *(Euderma maculatum)*, 58
- ■ **Hoary Bat** *(Lasiurus cinereus)*, 68
- ■ **Western Yellow Bat** *(Lasiurus xanthinus)*, 76
- ■ **Western Small-footed Bat** *(Myotis ciliolabrum)*, 84
- ■ **Fringed Bat** *(Myotis thysanodes)*, 100
- ▨ **Townsend's Big-eared Bat** *(Corynorhinus townsendii)*, 116
- ■ **Greater Mastiff Bat** *(Eumops perotis)*, 120
- ■ **Pocketed Free-tailed Bat** *(Nyctinomops femorosaccus)*, 126
- ■ **Big Free-tailed Bat** *(Nyctinomops macrotis)*, 128

Colorado

Common

- ■ **Pallid Bat** *(Antrozous pallidus)*, 54
- ■ **Big Brown Bat** *(Eptesicus fuscus)*, 56
- ▨ **Silver-haired Bat** *(Lasionycteris noctivagans)*, 62
- ▨ **Hoary Bat** *(Lasiurus cinereus)*, 68
- ■ **Western Small-footed Bat** *(Myotis ciliolabrum)*, 84
- ■ **Western Long-eared Bat** *(Myotis evotis)*, 86
- ■ **Little Brown Bat** *(Myotis lucifugus)*, 94
- ■ **Long-legged Bat** *(Myotis volans)*, 104
- ■ **Yuma Bat** *(Myotis yumanensis)*, 106
- ▨ **Western Pipistrelle Bat** *(Pipistrellus hesperus)*, 110

Uncommon

- ■ **Spotted Bat** *(Euderma maculatum)*, 58
- ▨ **Eastern Red Bat** *(Lasiurus borealis)*, 66
- ▨ **California Bat** *(Myotis californicus)*, 82
- ■ **Fringed Bat** *(Myotis thysanodes)*, 100
- ▨ **Eastern Pipistrelle Bat** *(Pipistrellus subflavus)*, 112
  Found in Colo. only once; considered an accidental
- ■ **Townsend's Big-eared Bat** *(Corynorhinus townsendii)*, 116
- ■ **Big Free-tailed Bat** *(Nyctinomops macrotis)*, 128
- ■ **Brazilian Free-tailed Bat** *(Tadarida brasiliensis)*, 130
  Cave Bat *(Myotis velifer)* and Allen's Big-eared Bat *(Idionycteris phyllotis)* are likely to reside in Colorado, but records have not been confirmed.

Connecticut

Common

- ■ **Big Brown Bat** *(Eptesicus fuscus)*, 56
- ■ **Little Brown Bat** *(Myotis lucifugus)*, 94
- ■ **Northern Long-eared Bat** *(Myotis septentrionalis)*, 96
- ▨ **Eastern Pipistrelle Bat** *(Pipistrellus subflavus)*, 112
  Common and widespread; however not often seen outside winter hibernaculum counts

Uncommon

- **Silver-haired Bat** *(Lasionycteris noctivagans)*, 62
- **Eastern Red Bat** *(Lasiurus borealis)*, 66
- **Hoary Bat** *(Lasiurus cinereus)*, 68
- **Eastern Small-footed Bat** *(Myotis leibii)*, 92
  Uncertain if it still is present in Conn.
- **Indiana Bat** *(Myotis sodalis)*, 98

Delaware

Common

- **Big Brown Bat** *(Eptesicus fuscus)*, 56
- **Silver-haired Bat** *(Lasionycteris noctivagans)*, 62
  Status unknown, but probably fairly common
- **Little Brown Bat** *(Myotis lucifugus)*, 94
- **Eastern Pipistrelle Bat** *(Pipistrellus subflavus)*, 112

Unknown Status

- **Eastern Red Bat** *(Lasiurus borealis)*, 66
- **Hoary Bat** *(Lasiurus cinereus)*, 68
- **Eastern Small-footed Bat** *(Myotis leibii)*, 92
- **Northern Long-eared Bat** *(Myotis septentrionalis)*, 96
- **Evening Bat** *(Nycticeius humeralis)*, 108

Florida

Common

- **Eastern Red Bat** *(Lasiurus borealis)*, 66
- **Northern Yellow Bat** *(Lasiurus intermedius)*, 72
- **Seminole Bat** *(Lasiurus seminolus)*, 74
- **Southeastern Bat** *(Myotis austroriparius)*, 80
- **Evening Bat** *(Nycticeius humeralis)*, 108
- **Eastern Pipistrelle Bat** *(Pipistrellus subflavus)*, 112
- **Brazilian Free-tailed Bat** *(Tadarida brasiliensis)*, 130

Uncommon

- **Jamaican Fruit-eating Bat** *(Artibeus jamaicensis)*, 52
- **Big Brown Bat** *(Eptesicus fuscus)*, 56
- **Silver-haired Bat** *(Lasionycteris noctivagans)*, 62
  Considered an accidental
- **Hoary Bat** *(Lasiurus cinereus)*, 68
- **Gray Bat** *(Myotis grisescens)*, 88
- **Little Brown Bat** *(Myotis lucifugus)*, 94
  Considered an accidental
- **Indiana Bat** *(Myotis sodalis)*, 98
- **Northern Long-eared Bat** *(Myotis septentrionalis)*, 96
  Considered an accidental
- **Rafinesque's Big-eared Bat** *(Corynorhinus rafinesquii)*, 114

- **Wagner's Mastiff Bat** *(Eumops glaucinus)*, 118
- **Pallas' Mastiff Bat** *(Molossus molossus)*, 124
  Uncommon throughout most of Fla., may be common in Keys

Georgia

Common

- **Big Brown Bat** *(Eptesicus fuscus)*, 56
- **Eastern Red Bat** *(Lasiurus borealis)*, 66
- **Seminole Bat** *(Lasiurus seminolus)*, 74
- **Northern Long-eared Bat** *(Myotis septentrionalis)*, 96
- **Evening Bat** *(Nycticeius humeralis)*, 108
- **Eastern Pipistrelle Bat** *(Pipistrellus subflavus)*, 112
- **Brazilian Free-tailed Bat** *(Tadarida brasiliensis)*, 130
  Common only in southern Georgia

Uncommon

- **Silver-haired Bat** *(Lasionycteris noctivagans)*, 62
- **Hoary Bat** *(Lasiurus cinereus)*, 68
- **Northern Yellow Bat** *(Lasiurus intermedius)*, 72
  Status unknown, but probably uncommon
- **Southeastern Bat** *(Myotis austroriparius)*, 80
- **Gray Bat** *(Myotis grisescens)*, 88
- **Eastern Small-footed Bat** *(Myotis leibii)*, 92

- **Little Brown Bat** *(Myotis lucifugus)*, 94
- **Indiana Bat** *(Myotis sodalis)*, 98
- **Rafinesque's Big-eared Bat** *(Corynorhinus rafinesquii)*, 114

Hawaii

Uncommon

- **Hoary Bat** *(Lasiurus cinereus)*, 68

Idaho

Common

- **Pallid Bat** *(Antrozous pallidus)*, 54
- **Big Brown Bat** *(Eptesicus fuscus)*, 56
- **Silver-haired Bat** *(Lasionycteris noctivagans)*, 62
- **Hoary Bat** *(Lasiurus cinereus)*, 68
- **California Bat** *(Myotis californicus)*, 82
- **Western Small-footed Bat** *(Myotis ciliolabrum)*, 84
- **Western Long-eared Bat** *(Myotis evotis)*, 86
  Moderately common to uncommon
- **Little Brown Bat** *(Myotis lucifugus)*, 94
- **Long-legged Bat** *(Myotis volans)*, 104
- **Yuma Bat** *(Myotis yumanensis)*, 106
- **Western Pipistrelle Bat** *(Pipistrellus hesperus)*, 110
- **Townsend's Big-eared Bat** *(Corynorhinus townsendii)*, 116

Uncommon

- **Spotted Bat** *(Euderma maculatum)*, 58
- **Fringed Bat** *(Myotis thysanodes)*, 100

Illinois

Common

- **Big Brown Bat** *(Eptesicus fuscus)*, 56
- **Silver-haired Bat** *(Lasionycteris noctivagans)*, 62
- **Eastern Red Bat** *(Lasiurus borealis)*, 66
- **Little Brown Bat** *(Myotis lucifugus)*, 94
- **Northern Long-eared Bat** *(Myotis septentrionalis)*, 96
- **Eastern Pipistrelle Bat** *(Pipistrellus subflavus)*, 112

Uncommon

- **Hoary Bat** *(Lasiurus cinereus)*, 68
- **Southeastern Bat** *(Myotis austroriparius)*, 80
- **Gray Bat** *(Myotis grisescens)*, 88
- **Indiana Bat** *(Myotis sodalis)*, 98
- **Evening Bat** *(Nycticeius humeralis)*, 108
- **Rafinesque's Big-eared Bat** *(Corynorhinus rafinesquii)*, 114

Indiana

Common

- **Big Brown Bat** *(Eptesicus fuscus)*, 56
- **Eastern Red Bat** *(Lasiurus borealis)*, 66
- **Little Brown Bat** *(Myotis lucifugus)*, 94
- **Northern Long-eared Bat** *(Myotis septentrionalis)*, 104
- **Eastern Pipistrelle Bat** *(Pipistrellus subflavus)*, 112

Uncommon

- **Silver-haired Bat** *(Lasionycteris noctivagans)*, 62
- **Hoary Bat** *(Lasiurus cinereus)*, 68
- **Southeastern Bat** *(Myotis austroriparius)*, 80
  Very rare, possibly extinct in the state
- **Gray Bat** *(Myotis grisescens)*, 88
- **Indiana Bat** *(Myotis sodalis)*, 98
- **Evening Bat** *(Nycticeius humeralis)*, 108
- **Rafinesque's Big-eared Bat** *(Corynorhinus rafinesquii)*, 114
  Has not been reported in Ind. for over 25 years

Iowa

Common

- **Big Brown Bat** *(Eptesicus fuscus)*, 56
- **Eastern Red Bat** *(Lasiurus borealis)*, 66
- **Little Brown Bat** *(Myotis lucifugus)*, 94

- **Northern Long-eared Bat** *(Myotis septentrionalis)*, 104
- **Eastern Pipistrelle Bat** *(Pipistrellus subflavus)*, 112

Uncommon

- **Silver-haired Bat** *(Lasionycteris noctivagans)*, 62
- **Hoary Bat** *(Lasiurus cinereus)*, 68
- **Indiana Bat** *(Myotis sodalis)*, 98
- **Evening Bat** *(Nycticeius humeralis)*, 108

Kansas

Common

- **Big Brown Bat** *(Eptesicus fuscus)*, 56
- **Silver-haired Bat** *(Lasionycteris noctivagans)*, 62
- **Eastern Red Bat** *(Lasiurus borealis)*, 66
- **Little Brown Bat** *(Myotis lucifugus)*, 94
- **Cave Bat** *(Myotis velifer)*, 102
- **Evening Bat** *(Nycticeius humeralis)*, 108
- **Eastern Pipistrelle Bat** *(Pipistrellus subflavus)*, 112
- **Townsend's Big-eared Bat** *(Corynorhinus townsendii)*, 116
- **Big Free-tailed Bat** *(Nyctinomops macrotis)*, 128
- **Brazilian Free-tailed Bat** *(Tadarida brasiliensis)*, 130

Uncommon

- **Pallid Bat** *(Antrozous pallidus)*, 54
- **Hoary Bat** *(Lasiurus cinereus)*, 68
- **Western Small-footed Bat** *(Myotis ciliolabrum)*, 84
- **Gray Bat** *(Myotis grisescens)*, 88

Kentucky

Common

- **Big Brown Bat** *(Eptesicus fuscus)*, 56
- **Eastern Red Bat** *(Lasiurus borealis)*, 66
- **Little Brown Bat** *(Myotis lucifugus)*, 94
- **Northern Long-eared Bat** *(Myotis septentrionalis)*, 96
- **Eastern Pipistrelle Bat** *(Pipistrellus subflavus)*, 112

Uncommon

- **Hoary Bat** *(Lasiurus cinereus)*, 68
- **Seminole Bat** *(Lasiurus seminolus)*, 74
  Considered an accidental
- **Southeastern Bat** *(Myotis austroriparius)*, 80
- **Gray Bat** *(Myotis grisescens)*, 88
- **Eastern Small-footed Bat** *(Myotis leibii)*, 92
- **Indiana Bat** *(Myotis sodalis)*, 98
- **Evening Bat** *(Nycticeius humeralis)*, 108

- **Rafinesque's Big-eared Bat** *(Corynorhinus rafinesquii)*, 114
- **Townsend's Big-eared Bat** *(Corynorhinus townsendii)*, 116
- **Brazilian Free-tailed Bat** *(Tadarida brasiliensis)*, 130
  Considered an accidental

Unknown Status

- **Silver-haired Bat** *(Lasionycteris noctivagans)*, 62
  Seems to occur only as a transient

Louisiana

Common

- **Big Brown Bat** *(Eptesicus fuscus)*, 56
- **Eastern Red Bat** *(Lasiurus borealis)*, 66
- **Seminole Bat** *(Lasiurus seminolus)*, 74
- **Evening Bat** *(Nycticeius humeralis)*, 108
- **Eastern Pipistrelle Bat** *(Pipistrellus subflavus)*, 112
- **Brazilian Free-tailed Bat** *(Tadarida brasiliensis)*, 130

Uncommon

- **Silver-haired Bat** *(Lasionycteris noctivagans)*, 62
- **Hoary Bat** *(Lasiurus cinereus)*, 68
- **Northern Yellow Bat** *(Lasiurus intermedius)*, 72
- **Southeastern Bat** *(Myotis austroriparius)*, 80
- **Rafinesque's Big-eared Bat** *(Corynorhinus rafinesquii)*, 114

Maine

Common

- **Big Brown Bat** *(Eptesicus fuscus)*, 56
- **Little Brown Bat** *(Myotis lucifugus)*, 94
- **Northern Long-eared Bat** *(Myotis septentrionalis)*, 96

Uncommon

- **Hoary Bat** *(Lasiurus cinereus)*, 68
- **Eastern Small-footed Bat** *(Myotis leibii)*, 92

Unknown Status

- **Silver-haired Bat** *(Lasionycteris noctivagans)*, 62
- **Eastern Red Bat** *(Lasiurus borealis)*, 66
- **Eastern Pipistrelle Bat** *(Pipistrellus subflavus)*, 112

Maryland

Common

- **Big Brown Bat** *(Eptesicus fuscus)*, 56
- **Eastern Red Bat** *(Lasiurus borealis)*, 66
- **Little Brown Bat** *(Myotis lucifugus)*, 94
- **Eastern Pipistrelle Bat** *(Pipistrellus subflavus)*, 112

Uncommon

- **Hoary Bat** *(Lasiurus cinereus)*, 68

- **Northern Yellow Bat** *(Lasiurus intermedius)*, 72
- **Eastern Small-footed Bat** *(Myotis leibii)*, 92
- **Northern Long-eared Bat** *(Myotis septentrionalis)*, 96
- **Indiana Bat** *(Myotis sodalis)*, 98

Unknown Status

- **Silver-haired Bat** *(Lasionycteris noctivagans)*, 62
- **Evening Bat** *(Nycticeius humeralis)*, 108

Massachusetts

Common

- **Big Brown Bat** *(Eptesicus fuscus)*, 56
- **Silver-haired Bat** *(Lasionycteris noctivagans)*, 62
- **Eastern Red Bat** *(Lasiurus borealis)*, 66
- **Little Brown Bat** *(Myotis lucifugus)*, 94
- **Northern Long-eared Bat** *(Myotis septentrionalis)*, 96

Uncommon

- **Hoary Bat** *(Lasiurus cinereus)*, 68
- **Eastern Small-footed Bat** *(Myotis leibii)*, 92
- **Eastern Pipistrelle Bat** *(Pipistrellus subflavus)*, 112

Michigan

Common

- **Big Brown Bat** *(Eptesicus fuscus)*, 56
- **Eastern Red Bat** *(Lasiurus borealis)*, 66
- **Little Brown Bat** *(Myotis lucifugus)*, 94
  Common in northern regions of state

Uncommon

- **Silver-haired Bat** *(Lasionycteris noctivagans)*, 62
- **Hoary Bat** *(Lasiurus cinereus)*, 68
- **Northern Long-eared Bat** *(Myotis septentrionalis)*, 96
- **Indiana Bat** *(Myotis sodalis)*, 98
- **Evening Bat** *(Nycticeius humeralis)*, 108
- **Eastern Pipistrelle Bat** *(Pipistrellus subflavus)*, 112

Minnesota

Common

- **Big Brown Bat** *(Eptesicus fuscus)*, 56
- **Silver-haired Bat** *(Lasionycteris noctivagans)*, 62
- **Little Brown Bat** *(Myotis lucifugus)*, 94

Uncommon

- **Eastern Red Bat** *(Lasiurus borealis)*, 66
  Populations seem to be declining

- Hoary Bat *(Lasiurus cinereus)*, 68
- Northern Long-eared Bat *(Myotis septentrionalis)*, 96
- Eastern Pipistrelle Bat *(Pipistrellus subflavus)*, 112

Mississippi

Common

- Big Brown Bat *(Eptesicus fuscus)*, 56
- Eastern Red Bat *(Lasiurus borealis)*, 66
- Seminole Bat *(Lasiurus seminolus)*, 74
- Evening Bat *(Nycticeius humeralis)*, 108
- Eastern Pipistrelle Bat *(Pipistrellus subflavus)*, 112
- Brazilian Free-tailed Bat *(Tadarida brasiliensis)*, 130

Uncommon

- Silver-haired Bat *(Lasionycteris noctivagans)*, 62
- Hoary Bat *(Lasiurus cinereus)*, 68
- Northern Yellow Bat *(Lasiurus intermedius)*, 72
  Status unknown, but probably uncommon
- Southeastern Bat *(Myotis austroriparius)*, 80
- Gray Bat *(Myotis grisescens)*, 88
- Little Brown Bat *(Myotis lucifugus)*, 94
- Northern Long-eared Bat *(Myotis septentrionalis)*, 96
- Indiana Bat *(Myotis sodalis)*, 98

Unknown Status

- Rafinesque's Big-eared Bat *(Corynorhinus rafinesquii)*, 114

Missouri

Common

- Big Brown Bat *(Eptesicus fuscus)*, 56
- Eastern Red Bat *(Lasiurus borealis)*, 66
- Little Brown Bat *(Myotis lucifugus)*, 94
- Eastern Pipistrelle Bat *(Pipistrellus subflavus)*, 112

Uncommon

- Silver-haired Bat *(Lasionycteris noctivagans)*, 62
- Hoary Bat *(Lasiurus cinereus)*, 68
- Gray Bat *(Myotis grisescens)*, 88
- Northern Long-eared Bat *(Myotis septentrionalis)*, 96
- Indiana Bat *(Myotis sodalis)*, 98
- Evening Bat *(Nycticeius humeralis)*, 108
  Found only in summer
- Rafinesque's Big-eared Bat *(Corynorhinus rafinesquii)*, 114
- Townsend's Big-eared Bat *(Corynorhinus townsendii)*, 116

Montana

Common

- Big Brown Bat *(Eptesicus fuscus)*, 56
- Silver-haired Bat *(Lasionycteris noctivagans)*, 62
- Hoary Bat *(Lasiurus cinereus)*, 68
- Western Small-footed Bat *(Myotis ciliolabrum)*, 84
- Western Long-eared Bat *(Myotis evotis)*, 86
  Widespread; common to uncommon
- Little Brown Bat *(Myotis lucifugus)*, 94
- Long-legged Bat *(Myotis volans)*, 104
- Yuma Bat *(Myotis yumanensis)*, 106
  Widespread; common to uncommon

Uncommon

- Pallid Bat *(Antrozous pallidus)*, 54
- Spotted Bat *(Euderma maculatum)*, 58
- Eastern Red Bat *(Lasiurus borealis)*, 66
- California Bat *(Myotis californicus)*, 82
- Northern Long-eared Bat *(Myotis septentrionalis)*, 96
- Fringed Bat *(Myotis thysanodes)*, 100
- Townsend's Big-eared Bat *(Corynorhinus townsendii)*, 116

Nebraska

Common

- Big Brown Bat *(Eptesicus fuscus)*, 56
- Silver-haired Bat *(Lasionycteris noctivagans)*, 62
- Eastern Red Bat *(Lasiurus borealis)*, 66
- Hoary Bat *(Lasiurus cinereus)*, 68
- Western Small-footed Bat *(Myotis ciliolabrum)*, 84
- Little Brown Bat *(Myotis lucifugus)*, 94
- Northern Long-eared Bat *(Myotis septentrionalis)*, 96

Uncommon

- Fringed Bat *(Myotis thysanodes)*, 100
- Long-legged Bat *(Myotis volans)*, 104
- Evening Bat *(Nycticeius humeralis)*, 108
- Eastern Pipistrelle Bat *(Pipistrellus subflavus)*, 112
- Townsend's Big-eared Bat *(Corynorhinus townsendii)*, 116
- Brazilian Free-tailed Bat *(Tadarida brasiliensis)*, 130

Nevada

Common

- Pallid Bat *(Antrozous pallidus)*, 54
- Big Brown Bat *(Eptesicus fuscus)*, 56

- **Silver-haired Bat** *(Lasionycteris noctivagans)*, 62
  Common seasonally in northern Nev., uncommon seasonally in southern Nev.
- **California Bat** *(Myotis californicus)*, 82
- **Western Small-footed Bat** *(Myotis ciliolabrum)*, 84
- **Western Long-eared Bat** *(Myotis evotis)*, 86
- **Little Brown Bat** *(Myotis lucifugus)*, 94
  Locally common in northern Nev.
- **Long-legged Bat** *(Myotis volans)*, 104
- **Yuma Bat** *(Myotis yumanensis)*, 106
  Common in N.W. Nev. and in south along major river drainages
- **Western Pipistrelle Bat** *(Pipistrellus hesperus)*, 110
- **Townsend's Big-eared Bat** *(Corynorhinus townsendii)*, 116
  Locally common
- **Big Free-tailed Bat** *(Nyctinomops macrotis)*, 128
  Locally and seasonally common in S.E. Nev.
- **Brazilian Free-tailed Bat** *(Tadarida brasiliensis)*, 130

Uncommon

- **California Leaf-nosed Bat** *(Macrotus californicus)*, 44
  Restricted to Clark County
- **Mexican Long-tongued Bat** *(Choeronycteris mexicana)*, 46
  Considered an accidental

- **Spotted Bat** *(Euderma maculatum)*, 58
  May be locally common seasonally
- **Allen's Big-eared Bat** *(Idionycteris phyllotis)*, 60
- **Western Red Bat** *(Lasiurus blossevillii)*, 64
- **Hoary Bat** *(Lasiurus cinereus)*, 68
- **Western Yellow Bat** *(Lasiurus xanthinus)*, 76
  Uncommon, but is common at one locality
- **Fringed Bat** *(Myotis thysanodes)*, 100
- **Cave Bat** *(Myotis velifer)*, 102
  Only one record ever found and not verified
- **Greater Mastiff Bat** *(Eumops perotis)*, 120
  Considered an accidental

New Hampshire

Common

- **Big Brown Bat** *(Eptesicus fuscus)*, 56
- **Little Brown Bat** *(Myotis lucifugus)*, 94
- **Northern Long-eared Bat** *(Myotis septentrionalis)*, 96

Uncommon

- **Eastern Small-footed Bat** *(Myotis leibii)*, 92
- **Indiana Bat** *(Myotis sodalis)*, 98
  Considered an accidental

New Jersey

New Mexico

- **Western Yellow Bat** *(Lasiurus xanthinus)*, 76
- **Little Brown Bat** *(Myotis lucifugus)*, 94
- **Fringed Bat** *(Myotis thysanodes)*, 100
- **Townsend's Big-eared Bat** *(Corynorhinus townsendii)*, 116
- **Greater Mastiff Bat** *(Eumops perotis)*, 120
- **Underwood's Mastiff Bat** *(Eumops underwoodi)*, 122
- **Pocketed Free-tailed Bat** *(Nyctinomops femorosaccus)*, 126
- **Big Free-tailed Bat** *(Nyctinomops macrotis)*, 128

New York

Common

- **Big Brown Bat** *(Eptesicus fuscus)*, 56
- **Little Brown Bat** *(Myotis lucifugus)*, 94
- **Northern Long-eared Bat** *(Myotis septentrionalis)*, 96

Uncommon

- **Eastern Small-footed Bat** *(Myotis leibii)*, 92
- **Indiana Bat** *(Myotis sodalis)*, 98
- **Eastern Pipistrelle Bat** *(Pipistrellus subflavus)*, 112

Unknown Status

- **Silver-haired Bat** *(Lasionycteris noctivagans)*, 62
- **Eastern Red Bat** *(Lasiurus borealis)*, 66
- **Hoary Bat** *(Lasiurus cinereus)*, 68

North Carolina

Common

- **Big Brown Bat** *(Eptesicus fuscus)*, 56
- **Eastern Red Bat** *(Lasiurus borealis)*, 66
- **Hoary Bat** *(Lasiurus cinereus)*, 68
  Common seasonally
- **Little Brown Bat** *(Myotis lucifugus)*, 94
- **Northern Long-eared Bat** *(Myotis septentrionalis)*, 96
- **Evening Bat** *(Nycticeius humeralis)*, 108
- **Eastern Pipistrelle Bat** *(Pipistrellus subflavus)*, 112
- **Brazilian Free-tailed Bat** *(Tadarida brasiliensis)*, 130

Uncommon

- **Silver-haired Bat** *(Lasionycteris noctivagans)*, 62
- **Northern Yellow Bat** *(Lasiurus intermedius)*, 72
  Only one record known
- **Southeastern Bat** *(Myotis austroriparius)*, 80
  Only one record known
- **Gray Bat** *(Myotis grisescens)*, 88
  Only one record known
- **Indiana Bat** *(Myotis sodalis)*, 98
- **Rafinesque's Big-eared Bat** *(Corynorhinus rafinesquii)*, 114
- **Townsend's Big-eared Bat** *(Corynorhinus townsendii)*, 116

Unknown Status

- **Seminole Bat** (*Lasiurus seminolus*), 74
- **Eastern Small-footed Bat** (*Myotis leibii*), 92

North Dakota

Common

- **Big Brown Bat** (*Eptesicus fuscus*), 56
- **Silver-haired Bat** (*Lasionycteris noctivagans*), 62
  Status unknown, but probably fairly common
- **Little Brown Bat** (*Myotis lucifugus*), 94
  Status unknown, but probably common

Uncommon

- **Hoary Bat** (*Lasiurus cinereus*), 68
  Status unknown, but probably uncommon
- **Western Small-footed Bat** (*Myotis ciliolabrum*), 84
- **Western Long-eared Bat** (*Myotis evotis*), 86
- **Long-legged Bat** (*Myotis volans*), 104

Unknown Status

- **Eastern Red Bat** (*Lasiurus borealis*), 66
- **Northern Long-eared Bat** (*Myotis septentrionalis*), 96

Ohio

Common

- **Big Brown Bat** (*Eptesicus fuscus*), 56
- **Eastern Red Bat** (*Lasiurus borealis*), 66
- **Little Brown Bat** (*Myotis lucifugus*), 94
- **Northern Long-eared Bat** (*Myotis septentrionalis*), 96
- **Eastern Pipistrelle Bat** (*Pipistrellus subflavus*), 112

Uncommon

- **Silver-haired Bat** (*Lasionycteris noctivagans*), 62
- **Hoary Bat** (*Lasiurus cinereus*), 68
- **Eastern Small-footed Bat** (*Myotis leibii*), 92
  Considered an accidental
- **Indiana Bat** (*Myotis sodalis*), 98
- **Evening Bat** (*Nycticeius humeralis*), 108
- **Rafinesque's Big-eared Bat** (*Corynorhinus rafinesquii*), 114

Oklahoma

Common

- **Big Brown Bat** (*Eptesicus fuscus*), 56
- **Eastern Red Bat** (*Lasiurus borealis*), 66
- **Hoary Bat** (*Lasiurus cinereus*), 68
- **Cave Bat** (*Myotis velifer*), 102

- **Evening Bat** *(Nycticeius humeralis)*, 108
- **Eastern Pipistrelle Bat** *(Pipistrellus subflavus)*, 112
- **Brazilian Free-tailed Bat** *(Tadarida brasiliensis)*, 130
  Locally common

Uncommon

- **Pallid Bat** *(Antrozous pallidus)*, 54
- **Silver-haired Bat** *(Lasionycteris noctivagans)*, 62
- **Seminole Bat** *(Lasiurus seminolus)*, 74
- **Southeastern Bat** *(Myotis austroriparius)*, 80
- **Western Small-footed Bat** *(Myotis ciliolabrum)*, 84
- **Gray Bat** *(Myotis grisescens)*, 88
- **Eastern Small-footed Bat** *(Myotis leibii)*, 92
- **Little Brown Bat** *(Myotis lucifugus)*, 94
- **Northern Long-eared Bat** *(Myotis septentrionalis)*, 96
- **Indiana Bat** *(Myotis sodalis)*, 98
- **Yuma Bat** *(Myotis yumanensis)*, 106
  Locally common in some areas
- **Western Pipistrelle Bat** *(Pipistrellus hesperus)*, 110
- **Rafinesque's Big-eared Bat** *(Corynorhinus rafinesquii)*, 114
- **Townsend's Big-eared Bat** *(Corynorhinus townsendii)*, 116
- **Big Free-tailed Bat** *(Nyctinomops macrotis)*, 128

Common

- **Pallid Bat** *(Antrozous pallidus)*, 54
  Mostly in central and eastern part of state
- **Big Brown Bat** *(Eptesicus fuscus)*, 56
- **California Bat** *(Myotis californicus)*, 82
- **Western Small-footed Bat** *(Myotis ciliolabrum)*, 84
- **Little Brown Bat** *(Myotis lucifugus)*, 94
- **Fringed Bat** *(Myotis thysanodes)*, 100
  Locally common
- **Long-legged Bat** *(Myotis volans)*, 104
  Locally common
- **Yuma Bat** *(Myotis yumanensis)*, 106
  Locally common in some riparian habitats
- **Western Pipistrelle Bat** *(Pipistrellus hesperus)*, 110
- **Brazilian Free-tailed Bat** *(Tadarida brasiliensis)*, 130
  Locally common

Uncommon

- **Spotted Bat** *(Euderma maculatum)*, 58
- **Silver-haired Bat** *(Lasionycteris noctivagans)*, 62
  Much habitat has been destroyed recently due to logging
- **Hoary Bat** *(Lasiurus cinereus)*, 68

- **Western Long-eared Bat** *(Myotis evotis)*, 86
- **Townsend's Big-eared Bat** *(Corynorhinus townsendii)*, 116

Pennsylvania

Common

- **Big Brown Bat** *(Eptesicus fuscus)*, 56
- **Silver-haired Bat** *(Lasionycteris noctivagans)*, 62
- **Eastern Red Bat** *(Lasiurus borealis)*, 66
- **Little Brown Bat** *(Myotis lucifugus)*, 94
- **Eastern Pipistrelle Bat** *(Pipistrellus subflavus)*, 112

Uncommon

- **Hoary Bat** *(Lasiurus cinereus)*, 68
- **Seminole Bat** *(Lasiurus seminolus)*, 74
- **Eastern Small-footed Bat** *(Myotis leibii)*, 92
- **Northern Long-eared Bat** *(Myotis septentrionalis)*, 96
- **Indiana Bat** *(Myotis sodalis)*, 98
- **Evening Bat** *(Nycticeius humeralis)*, 108

Rhode Island

Common

- **Big Brown Bat** *(Eptesicus fuscus)*, 56
- **Eastern Red Bat** *(Lasiurus borealis)*, 66
- **Little Brown Bat** *(Myotis lucifugus)*, 94

Uncommon

- **Silver-haired Bat** *(Lasionycteris noctivagans)*, 62
- **Hoary Bat** *(Lasiurus cinereus)*, 68
- **Northern Long-eared Bat** *(Myotis septentrionalis)*, 96
- **Eastern Pipistrelle Bat** *(Pipistrellus subflavus)*, 112

South Carolina

Common

- **Big Brown Bat** *(Eptesicus fuscus)*, 56
- **Silver-haired Bat** *(Lasionycteris noctivagans)*, 62
  Common during winter months
- **Eastern Red Bat** *(Lasiurus borealis)*, 66
- **Seminole Bat** *(Lasiurus seminolus)*, 74
- **Evening Bat** *(Nycticeius humeralis)*, 108
- **Eastern Pipistrelle Bat** *(Pipistrellus subflavus)*, 112
- **Brazilian Free-tailed Bat** *(Tadarida brasiliensis)*, 130

Uncommon

- **Southeastern Bat** *(Myotis austroriparius)*, 80
- **Eastern Small-footed Bat** *(Myotis leibii)*, 92
- **Little Brown Bat** *(Myotis lucifugus)*, 94
- **Northern Long-eared Bat** *(Myotis septentrionalis)*, 96
- **Indiana Bat** *(Myotis sodalis)*, 98

STATE LISTS

■ **Rafinesque's Big-eared Bat** *(Corynorhinus rafinesquii)*, 114
■ **Big Free-tailed Bat** *(Nyctinomops macrotis)*, 128
   Considered an accidental

Unknown Status
■ **Hoary Bat** *(Lasiurus cinereus)*, 68
■ **Northern Yellow Bat** *(Lasiurus intermedius)*, 72

South Dakota

Common
■ **Big Brown Bat** *(Eptesicus fuscus)*, 56
■ **Silver-haired Bat** *(Lasionycteris noctivagans)*, 62
■ **Eastern Red Bat** *(Lasiurus borealis)*, 66
■ **Hoary Bat** *(Lasiurus cinereus)*, 68
   Fairly common in forested areas
■ **Little Brown Bat** *(Myotis lucifugus)*, 94

Uncommon
■ **Western Small-footed Bat** *(Myotis ciliolabrum)*, 84
■ **Western Long-eared Bat** *(Myotis evotis)*, 86
■ **Northern Long-eared Bat** *(Myotis septentrionalis)*, 96
■ **Fringed Bat** *(Myotis thysanodes)*, 100
■ **Long-legged Bat** *(Myotis volans)*, 104

■ **Evening Bat** *(Nycticeius humeralis)*, 108
   Reports of its presence in state are unconfirmed
■ **Townsend's Big-eared Bat** *(Corynorhinus townsendii)*, 116

Tennessee

Common
■ **Big Brown Bat** *(Eptesicus fuscus)*, 56
■ **Silver-haired Bat** *(Lasionycteris noctivagans)*, 62
■ **Eastern Red Bat** *(Lasiurus borealis)*, 66
■ **Hoary Bat** *(Lasiurus cinereus)*, 68
■ **Little Brown Bat** *(Myotis lucifugus)*, 94
■ **Northern Long-eared Bat** *(Myotis septentrionalis)*, 96
■ **Evening Bat** *(Nycticeius humeralis)*, 108
■ **Eastern Pipistrelle Bat** *(Pipistrellus subflavus)*, 112

Uncommon
■ **Seminole Bat** *(Lasiurus seminolus)*, 74
■ **Southeastern Bat** *(Myotis austroriparius)*, 80
■ **Gray Bat** *(Myotis grisescens)*, 88
■ **Eastern Small-footed Bat** *(Myotis leibii)*, 92
■ **Indiana Bat** *(Myotis sodalis)*, 98
■ **Rafinesque's Big-eared Bat** *(Corynorhinus rafinesquii)*, 114

Texas

Common

- **Pallid Bat** (*Antrozous pallidus*), 54
- **Big Brown Bat** (*Eptesicus fuscus*), 56
- **Silver-haired Bat** (*Lasionycteris noctivagans*), 62
- **Eastern Red Bat** (*Lasiurus borealis*), 66
- **Seminole Bat** (*Lasiurus seminolus*), 74
- **California Bat** (*Myotis californicus*), 44
- **Fringed Bat** (*Myotis thysanodes*), 100
- **Cave Bat** (*Myotis velifer*), 102
- **Yuma Bat** (*Myotis yumanensis*), 106
- **Evening Bat** (*Nycticeius humeralis*), 108
- **Western Pipistrelle Bat** (*Pipistrellus hesperus*), 110
- **Eastern Pipistrelle Bat** (*Pipistrellus subflavus*), 112
- **Brazilian Free-tailed Bat** (*Tadarida brasiliensis*), 130

Uncommon

- **Ghost-faced Bat** (*Mormoops megalophylla*), 42
- **Mexican Long-tongued Bat** (*Choeronycteris mexicana*), 46
- **Greater Long-nosed Bat** (*Leptonycteris nivalis*), 50
  Can be found in Brewster and Presidio Counties in Trans-Pecos
- **Spotted Bat** (*Euderma maculatum*), 58

- **Western Red Bat** (*Lasiurus blossevillii*), 64
  Considered an accidental
- **Hoary Bat** (*Lasiurus cinereus*), 68
- **Southern Yellow Bat** (*Lasiurus ega*), 70
- **Northern Yellow Bat** (*Lasiurus intermedius*), 72
- **Southeastern Bat** (*Myotis austroriparius*), 80
- **Western Small-footed Bat** (*Myotis ciliolabrum*), 84
- **Little Brown Bat** (*Myotis lucifugus*), 94
  Only one record known
- **Northern Long-eared Bat** (*Myotis septentrionalis*), 96
  Only one record known
- **Long-legged Bat** (*Myotis volans*), 104
- **Rafinesque's Big-eared Bat** (*Corynorhinus rafinesquii*), 114
- **Townsend's Big-eared Bat** (*Corynorhinus townsendii*), 116
- **Greater Mastiff Bat** (*Eumops perotis*), 120
- **Pocketed Free-tailed Bat** (*Nyctinomops femorosaccus*), 126
- **Big Free-tailed Bat** (*Nyctinomops macrotis*), 128

Utah

Common

- **Pallid Bat** (*Antrozous pallidus*), 54
- **Big Brown Bat** (*Eptesicus fuscus*), 56
- **Silver-haired Bat** (*Lasionycteris noctivagans*), 62

- **Western Red Bat** *(Lasiurus blossevillii)*, 64
- **California Bat** *(Myotis californicus)*, 82
- **Western Long-eared Bat** *(Myotis evotis)*, 86
- **Little Brown Bat** *(Myotis lucifugus)*, 94
- **Long-legged Bat** *(Myotis volans)*, 104
- **Western Pipistrelle Bat** *(Pipistrellus hesperus)*, 110
- **Townsend's Big-eared Bat** *(Corynorhinus townsendii)*, 116
- **Brazilian Free-tailed Bat** *(Tadarida brasiliensis)*, 130

Uncommon

- **Spotted Bat** *(Euderma maculatum)*, 58
- **Allen's Big-eared Bat** *(Idionycteris phyllotis)*, 60
- **Hoary Bat** *(Lasiurus cinereus)*, 68
- **Western Small-footed Bat** *(Myotis ciliolabrum)*, 84
- **Fringed Bat** *(Myotis thysanodes)*, 100
- **Yuma Bat** *(Myotis yumanensis)*, 106
- **Big Free-tailed Bat** *(Nyctinomops macrotis)*, 128

Vermont

Common

- **Big Brown Bat** *(Eptesicus fuscus)*, 56
- **Eastern Red Bat** *(Lasiurus borealis)*, 66
- **Hoary Bat** *(Lasiurus cinereus)*, 68
- **Little Brown Bat** *(Myotis lucifugus)*, 94

Uncommon

- **Silver-haired Bat** *(Lasionycteris noctivagans)*, 62
- **Eastern Small-footed Bat** *(Myotis leibii)*, 92
- **Northern Long-eared Bat** *(Myotis septentrionalis)*, 96
- **Indiana Bat** *(Myotis sodalis)*, 98
- **Eastern Pipistrelle Bat** *(Pipistrellus subflavus)*, 112

Virginia

Common

- **Big Brown Bat** *(Eptesicus fuscus)*, 56
- **Eastern Red Bat** *(Lasiurus borealis)*, 66
- **Little Brown Bat** *(Myotis lucifugus)*, 94
- **Evening Bat** *(Nycticeius humeralis)*, 108
  Common in lower elevations
- **Eastern Pipistrelle Bat** *(Pipistrellus subflavus)*, 112

Uncommon

- **Silver-haired Bat** *(Lasionycteris noctivagans)*, 62
  Found only in winter months
- **Hoary Bat** *(Lasiurus cinereus)*, 68
- **Northern Yellow Bat** *(Lasiurus intermedius)*, 72
- **Seminole Bat** *(Lasiurus seminolus)*, 74
  Considered an accidental

- **Southeastern Bat** *(Myotis austroriparius)*, 80
- **Gray Bat** *(Myotis grisescens)*, 88
- **Eastern Small-footed Bat** *(Myotis leibii)*, 92
- **Northern Long-eared Bat** *(Myotis septentrionalis)*, 96
- **Indiana Bat** *(Myotis sodalis)*, 98
- **Rafinesque's Big-eared Bat** *(Corynorhinus rafinesquii)*, 114
- **Townsend's Big-eared Bat** *(Corynorhinus townsendii)*, 116

Washington

Common

- **Big Brown Bat** *(Eptesicus fuscus)*, 56
- **Silver-haired Bat** *(Lasionycteris noctivagans)*, 62
- **Hoary Bat** *(Lasiurus cinereus)*, 68
- **California Bat** *(Myotis californicus)*, 82
- **Western Long-eared Bat** *(Myotis evotis)*, 86
  Moderately common to uncommon
- **Keen's Bat** *(Myotis keenii)*, 90
- **Little Brown Bat** *(Myotis lucifugus)*, 94
- **Long-legged Bat** *(Myotis volans)*, 104
- **Yuma Bat** *(Myotis yumanensis)*, 106
- **Townsend's Big-eared Bat** *(Corynorhinus townsendii)*, 116

Uncommon

- **Pallid Bat** *(Antrozous pallidus)*, 54
- **Spotted Bat** *(Euderma maculatum)*, 58
- **Western Small-footed Bat** *(Myotis ciliolabrum)*, 84
- **Fringed Bat** *(Myotis thysanodes)*, 100
- **Western Pipistrelle Bat** *(Pipistrellus hesperus)*, 110

Unknown Status

- **Western Red Bat** *(Lasiurus blossevillii)*, 64

West Virginia

Common

- **Big Brown Bat** *(Eptesicus fuscus)*, 56
- **Eastern Red Bat** *(Lasiurus borealis)*, 66
- **Little Brown Bat** *(Myotis lucifugus)*, 94
- **Northern Long-eared Bat** *(Myotis septentrionalis)*, 96
- **Eastern Pipistrelle Bat** *(Pipistrellus subflavus)*, 112

Uncommon

- **Silver-haired Bat** *(Lasionycteris noctivagans)*, 62
- **Hoary Bat** *(Lasiurus cinereus)*, 68
- **Gray Bat** *(Myotis grisescens)*, 88
- **Eastern Small-footed Bat** *(Myotis leibii)*, 92

- **Indiana Bat** *(Myotis sodalis)*, 98
- **Evening Bat** *(Nycticeius humeralis)*, 108
- **Rafinesque's Big-eared Bat** *(Corynorhinus rafinesquii)*, 114
- **Townsend's Big-eared Bat** *(Corynorhinus townsendii)*, 116

Wisconsin

Common

- **Big Brown Bat** *(Eptesicus fuscus)*, 56
- **Silver-haired Bat** *(Lasionycteris noctivagans)*, 62
- **Eastern Red Bat** *(Lasiurus borealis)*, 66
  Common in summer months
- **Hoary Bat** *(Lasiurus cinereus)*, 68
  Common in summer months
- **Little Brown Bat** *(Myotis lucifugus)*, 94
- **Northern Long-eared Bat** *(Myotis septentrionalis)*, 96
- **Eastern Pipistrelle Bat** *(Pipistrellus subflavus)*, 112

Uncommon

- **Indiana Bat** *(Myotis sodalis)*, 98
  Only one record known; considered an accidental

Wyoming

Common

- **Pallid Bat** *(Antrozous pallidus)*, 54
- **Big Brown Bat** *(Eptesicus fuscus)*, 56
- **Spotted Bat** *(Euderma maculatum)*, 58
- **Silver-haired Bat** *(Lasionycteris noctivagans)*, 62
- **Hoary Bat** *(Lasiurus cinereus)*, 68
- **Western Small-footed Bat** *(Myotis ciliolabrum)*, 84
- **Western Long-eared Bat** *(Myotis evotis)*, 86
- **Little Brown Bat** *(Myotis lucifugus)*, 94
- **Northern Long-eared Bat** *(Myotis septentrionalis)*, 96
- **Fringed Bat** *(Myotis thysanodes)*, 100
- **Long-legged Bat** *(Myotis volans)*, 104
- **Townsend's Big-eared Bat** *(Corynorhinus townsendii)*, 116

Uncommon

- **Eastern Red Bat** *(Lasiurus borealis)*, 66
- **California Bat** *(Myotis californicus)*, 82
- **Yuma Bat** *(Myotis yumanensis)*, 106
- **Eastern Pipistrelle Bat** *(Pipistrellus subflavus)*, 112
- **Big Free-tailed Bat** *(Nyctinomops macrotis)*, 128
- **Brazilian Free-tailed Bat** *(Tadarida brasiliensis)*, 130

CANADA

Alberta

Common

- Big Brown Bat *(Eptesicus fuscus)*, 56
- Silver-haired Bat *(Lasionycteris noctivagans)*, 62
- Hoary Bat *(Lasiurus cinereus)*, 68
- Western Long-eared Bat *(Myotis evotis)*, 86
- Little Brown Bat *(Myotis lucifugus)*, 94

Uncommon

- Eastern Red Bat *(Lasiurus borealis)*, 66
- Western Small-footed Bat *(Myotis ciliolabrum)*, 84
- Northern Long-eared Bat *(Myotis septentrionalis)*, 96
- Long-legged Bat *(Myotis volans)*, 104

British Columbia

Common

- Big Brown Bat *(Eptesicus fuscus)*, 56
- Silver-haired Bat *(Lasionycteris noctivagans)*, 62
- California Bat *(Myotis californicus)*, 82
- Western Long-eared Bat *(Myotis evotis)*, 86
  Moderately common to uncommon
- Little Brown Bat *(Myotis lucifugus)*, 94
- Yuma Bat *(Myotis yumanensis)*, 106

Uncommon

- Pallid Bat *(Antrozous pallidus)*, 54
- Spotted Bat *(Euderma maculatum)*, 58
- Western Red Bat *(Lasiurus blossevillii)*, 64
- Hoary Bat *(Lasiurus cinereus)*, 68
- Western Small-footed Bat *(Myotis ciliolabrum)*, 84
- Keen's Bat *(Myotis keenii)*, 90
- Northern Long-eared Bat *(Myotis septentrionalis)*, 96
- Fringed Bat *(Myotis thysanodes)*, 100
- Long-legged Bat *(Myotis volans)*, 104
- Townsend's Big-eared Bat *(Corynorhinus townsendii)*, 116

Manitoba

Common

- Big Brown Bat *(Eptesicus fuscus)*, 56
- Silver-haired Bat *(Lasionycteris noctivagans)*, 62
- Hoary Bat *(Lasiurus cinereus)*, 68
- Little Brown Bat *(Myotis lucifugus)*, 94
- Northern Long-eared Bat *(Myotis septentrionalis)*, 96

Uncommon

- Eastern Red Bat *(Lasiurus borealis)*, 66

New Brunswick

Common

- ■ **Little Brown Bat** *(Myotis lucifugus)*, 94
- ■ **Northern Long-eared Bat** *(Myotis septentrionalis)*, 96

Uncommon

- ■ **Big Brown Bat** *(Eptesicus fuscus)*, 56
- ■ **Silver-haired Bat** *(Lasionycteris noctivagans)*, 62
- ■ **Eastern Red Bat** *(Lasiurus borealis)*, 66
- ■ **Hoary Bat** *(Lasiurus cinereus)*, 68
- ■ **Eastern Pipistrelle Bat** *(Pipistrellus subflavus)*, 112

Newfoundland

Common

- ■ **Little Brown Bat** *(Myotis lucifugus)*, 94
- ■ **Northern Long-eared Bat** *(Myotis septentrionalis)*, 96

Northwest Territories

Common

- ■ **Hoary Bat** *(Lasiurus cinereus)*, 68
- ■ **Little Brown Bat** *(Myotis lucifugus)*, 94
- ■ **Northern Long-eared Bat** *(Myotis septentrionalis)*, 96

Nova Scotia

Common

- ■ **Little Brown Bat** *(Myotis lucifugus)*, 94
- ■ **Northern Long-eared Bat** *(Myotis septentrionalis)*, 96

Uncommon

- ■ **Big Brown Bat** *(Eptesicus fuscus)*, 56
- ■ **Silver-haired Bat** *(Lasionycteris noctivagans)*, 62
- ■ **Eastern Red Bat** *(Lasiurus borealis)*, 66
- ■ **Hoary Bat** *(Lasiurus cinereus)*, 68
- ■ **Eastern Pipistrelle Bat** *(Pipistrellus subflavus)*, 112

Nunavut Territory

No confirmed records of bats in Nunavut Territory

Ontario

Common

- ■ **Big Brown Bat** *(Eptesicus fuscus)*, 56
- ■ **Eastern Red Bat** *(Lasiurus borealis)*, 66
  Common especially in woodland areas
- ■ **Little Brown Bat** *(Myotis lucifugus)*, 94

Uncommon

- **Silver-haired Bat** *(Lasionycteris noctivagans)*, 62
- **Hoary Bat** *(Lasiurus cinereus)*, 68
- **Northern Long-eared Bat** *(Myotis septentrionalis)*, 96
- **Evening Bat** *(Nycticeius humeralis)*, 108
  Considered an accidental
- **Eastern Pipistrelle Bat** *(Pipistrellus subflavus)*, 112

Prince Edward Island

Common

- **Little Brown Bat** *(Myotis lucifugus)*, 94
- **Northern Long-eared Bat** *(Myotis septentrionalis)*, 96

Quebec

Common

- **Big Brown Bat** *(Eptesicus fuscus)*, 56
- **Silver-haired Bat** *(Lasionycteris noctivagans)*, 62
- **Eastern Red Bat** *(Lasiurus borealis)*, 66
- **Northern Long-eared Bat** *(Myotis septentrionalis)*, 96

Uncommon

- **Hoary Bat** *(Lasiurus cinereus)*, 68
- **Eastern Small-footed Bat** *(Myotis leibii)*, 92

Unknown Status

- **Eastern Pipistrelle Bat** *(Pipistrellus subflavus)*, 112

Saskatchewan

Common

- **Big Brown Bat** *(Eptesicus fuscus)*, 56
- **Silver-haired Bat** *(Lasionycteris noctivagans)*, 62
- **Hoary Bat** *(Lasiurus cinereus)*, 68
- **Little Brown Bat** *(Myotis lucifugus)*, 94
- **Northern Long-eared Bat** *(Myotis septentrionalis)*, 96

Uncommon

- **Eastern Red Bat** *(Lasiurus borealis)*, 66
- **Western Small-footed Bat** *(Myotis ciliolabrum)*, 84
- **Western Long-eared Bat** *(Myotis evotis)*, 86

Yukon Territory

Common

- **Little Brown Bat** *(Myotis lucifugus)*, 94
- **Northern Long-eared Bat** *(Myotis septentrionalis)*, 96

Uncommon

- **Long-legged Bat** *(Myotis volans)*, 104

Parts of a Bat

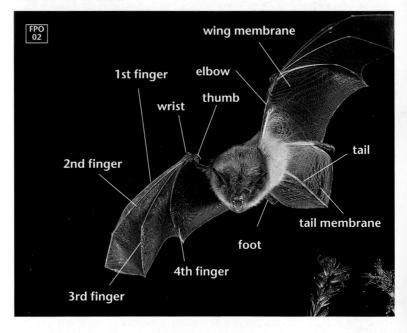

FPO 02

wing membrane
elbow
1st finger
thumb
wrist
2nd finger
tail
tail membrane
foot
4th finger
3rd finger

Guide to State Lists

Turn to the pages edged with the red bar to find a listing of the bats found in each U.S. state and Canadian province or territory. This listing will help you determine which bats you are likely to see where you are.